EXTRAORDINARY
STORIES

Linda Maron

LONGMEADOW PRESS

If you have been involved in a rescue or 9-1-1
experience that might be appropriate for
"Rescue 911", please contact Rae Chandler c/o:
Rescue 911, CBS-TV, Los Angeles, CA 90036 or call
1 (800) 678-4276.

Published by Longmeadow Press, 201 High Ridge Road, Stamford, CT
06904. All rights reserved. No part of this book may be reproduced
or utilized in any form or by any means, electronic or mechanical,
including photocopying, recording or by any information storage and
retrieval system, without permission in writing from the Publisher.

Jacket design by Lisa Amoroso
Interior design by Lisa Amoroso

Library of Congress Cataloging-in-Publication Data

Maron, Linda.
Rescue 911 : extraordinary stories / Linda Maron. – 1st ed.
p. cm.
ISBN 0-681-45256-0
1. Rescue work – United States – Case studies.   2. Emergency medical
services – United States – Case studies.   3. Rescues – United States.
I. Title
RA645.5.M375   1993
362.1'8–dc20   93-26636
                                                                             CIP
ISBN: 0–681–45256–0
Printed in United States of America
First Edition
0 9 8 7 6 5 4 3 2 1

Foreword

How do you account for the success of *Rescue 911*? I'm asked this question many times and the answer is not obvious. There are many other reality shows, some using footage that we would dearly love to have and for one reason or another haven't been able to get. Other shows do re-creations equally as well as we do, and, although I have less objectivity concerning the popularity of the various hosts, how much difference does a personality make? So the question remains—why is *Rescue 911* such a popular show in the national ratings?

The answer, I believe, is that we tug at the heartstrings of the audience. The chords we play resonate in some universal way. People want to see heroes at this time more than ever. They've had enough of greed and corruption, of cowardice and fear. They want to look at people who have instinctively reacted in a more noble fashion, people who by their actions demonstrate that there is a finer spirit in human beings than what we see on daily TV. It's a simple answer–man can be noble. I, for one, watch the show every week.

William Shatner

Introduction

Sirens . . . emergency vehicles rushing to the scene of a life and death emergency . . . heroic men and women risking their lives to save the lives of strangers—these are the images portrayed every Tuesday night on *Rescue 911*, the hit CBS television series now in its fifth season and watched by more than twenty million viewers across America, and by millions more in sixty countries around the world.

The 81 stories featured in this book were selected by the *Rescue 911* staff from more than 450 that we have produced. The same excitement, heroism, and concern for others that you experience watching our series can be experienced again reading this book— this time without commercial interruptions! These *Rescue 911* stories show dozens of truly remarkable men, women, and children who acted with courage and strength when called upon to rescue another in need.

To date, *Rescue 911*, as a direct result of life-saving techniques and procedures shown on our program, has been responsible for helping to save the lives of more than two hundred men, women, and children. While everyone who works on *Rescue 911* is extremely proud of the contribution our series is making to prevent tragedies, all of us know that there's no substitute for enrolling in a first-aid/CPR class offered by the American Red Cross or the American Heart Association. At any time, you or your child could be called upon to save the life of a loved one or a stranger.

Writing this introduction gives me the opportunity to thank the more than one hundred talented and hardworking people who make *Rescue 911*. I especially want to acknowledge Jim Milio, our Producer and Director, and Stephanie Siegel, our Footage Supervisor, for their contributions to this book. I would also like to thank our colleagues at CBS Entertainment Productions for their continued support; and the CBS television network for airing our series since its premiere in 1989.

Arnold Shapiro, Executive Producer, *Rescue 911*
Hollywood, California
June 1993

Arlington 9-1-1:
From the Pilot Episode

Eight-year-old Laura Hollingsworth thought she was dreaming when she opened her eyes and saw a man standing over the bed with a gun pointed at her. It was in Arlington, Texas, shortly before dawn on December 14, 1988, and Laura and her father, Dale, were asleep in Dale's bed. Laura's brother, fourteen-year-old Buddy, was asleep down the hall. Unfortunately, Laura wasn't dreaming.

The intruder woke up Dale, forced him out of the room at gunpoint, and made him cut the phone wires. Laura lay in bed, terrified, and heard the intruder yelling to an accomplice. Laura screamed for her father, and when he didn't respond, she ran into the living room and found him on the floor, scuffling with the robber. Laura raced to the phone in the kitchen, where luckily the wires had not yet been cut. She dialed 9-1-1.

"There's somebody in my house!" Laura's voice quivered in fear. "He broke in with a gun. He's threatening—"

"I have police officers on the way," responded Arlington Police Department Dispatcher Valerie Nelson, who heard the fighting in the background. "Stay on the phone."

As police officers sped to the scene, Nelson questioned Laura, but between Laura's panic and the commotion in the background, Nelson was unable to make sense out of what was happening.

"There's a gun and a knife," screamed Laura in tears.

"Who has the knife?" asked Nelson.

"My brother! He was going to kill my father if he didn't tell him where his wallet was!"

Laura of course meant *the robber* was going to kill her father, but it came across to Nelson as if Laura's brother and father were fighting.

"I think he's got the guy!"

"Who's got the guy?" asked Nelson.

"My dad."

"Your father has—"

"Buddy!" Laura let out a blood-curdling scream.

"What's going on?"

"Oh God! Buddy!" cried Laura hysterically. "Please!"

Nelson heard a shot fired.

"He killed the guy!" cried Laura.

"Your father shot the burglar?" asked Nelson.

"No, Buddy did."

"Your brother?"

"He shot my dad, too!"

As the confusion mounted, police officers responded to what they thought was a domestic disturbance. Nelson told Laura to open the door for the officers who stood outside with guns drawn. But Laura didn't trust the dispatcher, and Buddy thought the police outside might really be the robber's accomplices. The officers could see Buddy's silhouette inside the house. He was holding a rifle, but officers didn't realize he was a scared fourteen-year-old protecting himself and his little sister.

"I need your help," Nelson told Laura sternly. "It's the police outside. Can you go to the door?"

"No," replied Laura.

"It's the good people outside!" urged Nelson. "Tell Buddy to come to the phone."

Nelson desperately explained to Buddy that the police were afraid of him, that he had to come out before rescuers could safely get inside to help his wounded father. Finally Buddy dropped the rifle and came out, followed by Laura. Officers handcuffed Buddy, unaware that he'd just saved his father's life by killing the robber with his father's hunting rifle.

Buddy was quickly released from custody. Dale recovered from the stab wounds inflicted by the eighteen-year-old robber. But the accomplice was never found. Dale and his family did not spend another night in their house.

"Ugh, what a call," recalls dispatcher Nelson. "If I had those calls everyday, I could not do this job. You realize how close they came to losing it all."

"I literally didn't own anything of value that I would miss more than a half-minute," says Dale sadly. "I get upset thinking about that. I got all cut up, but, Jeez, you know, I didn't want my kid to kill somebody."

"For the rest of my life I'll always remember the day this happened," says Buddy. "It'll depend on the situation at the time if I'll feel happy we're alive, or I'll feel sad that it happened."

Nelson feels that the key to the successful outcome was Laura. Despite her terror, she had the presence of mind to call 9-1-1 and managed to remain calm and relay events over the phone.

Says Dale, "If Laura hadn't called 9-1-1, if Buddy hadn't intervened, if the dispatcher hadn't known how to do her job, I would be dead. And I feel my children would be, too."

Player Down

The Bulldogs' football game on November 7, 1986, was the last and biggest of the season for MacKenzie Phillips and his teammates of Springdale High School in Springdale, Arkansas. The annual "Dogfight" against archrival Fayetteville High School was well into its third quarter when MacKenzie, a promising seventeen-year-old defense lineman, didn't get up off the field after a play. He lay on the ground until his teammates came over to check on him. MacKenzie's eyes were rolled back, his face was blue, and he was completely unresponsive.

Team physicians ran out on the field and determined that MacKenzie was without breath or pulse. Surrounded by a stunned circle of players, doctors immediately began to administer CPR. They summoned rescuers, including paramedic Dave Creek and EMT John Baker, who were standing by on the field.

MacKenzie's parents, Betsy and Loyd Phillips, who had been watching from the bleachers, knew the situation was serious when Loyd was allowed on the playing field, which wasn't the norm following an injury. Then Loyd heard people say MacKenzie wasn't breathing. Once the crowd became aware that he had collapsed from cardiac arrest, the arena fell totally silent. Players knelt and prayed as EMT Baker took over chest compressions.

"It was uncanny," recalls Baker. "You could have heard a pin drop in the stadium."

MacKenzie wasn't responding to CPR. But suddenly, somebody gave a thumb's up and the crowd burst into cheers, thinking the worst was over. Their elation was short-lived, however, because the victory signal turned out to be a false alarm. It was now fifteen minutes since MacKenzie had collapsed. Paramedics had administered drugs and shocked his heart, but he had not regained a pulse. His parents stood by helplessly and watched.

"I was afraid of failure," recalls paramedic Creek. "Afraid that in front of all these people, that if we didn't get him resuscitated, it's going to be very bad. We all became very emotionally attached to the situation."

"What you end up doing is praying," says Baker, who got a hernia as a result of performing CPR that day. "You're praying, please don't let him die, please don't let him die."

MacKenzie was transported to the hospital, where additional rounds of medication to stimulate his heart and shocks to convert his heartbeat to a regular rhythm proved unsuccessful. Betsy held her son's hand while doctors continued to perform CPR. She squeezed MacKenzie's hand and whispered into his ear, "MacKenzie, you can make it. I love you."

Three minutes later, after one additional dose of medication, MacKenzie's heart spontaneously regained its regular rhythm and his blood pressure returned to normal. His heart had been stopped for twenty-five minutes, but miraculously, he suffered neither brain damage nor other permanent effects as a result.

Betsy believes that the extraordinary efforts of everybody involved in MacKenzie's rescue made the difference between life and death.

"I couldn't say thank you enough. There's no way you can ever thank anybody for your child's life."

The work of rescuers also touched MacKenzie.

"It's hard to really express in words my deep gratitude. A man who would give CPR for so long that it would cause himself to have a hernia—it's really amazing that he could care so much about a person he didn't even know."

MacKenzie received medication to control the asthma that caused his cardiac arrest, and in his first year of college, he was able to follow in his father's footsteps and start as tackle for the University of Arkansas Razorbacks.

MacKenzie and his father consider themselves best friends and feel lucky they still have each other today.

"The paramedics did great," says Loyd. "The doctors, the hospital, did great. But something like that—it's unbelievable. It's just a miracle."

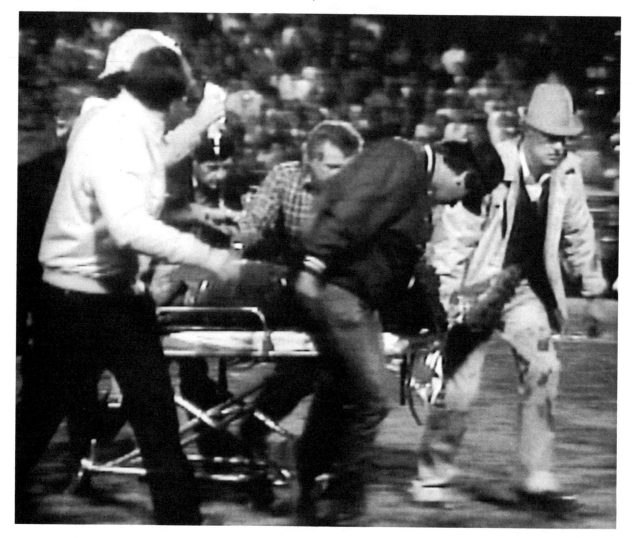

9-1-1 Rifle

Danny Jr. and Georgia Troy, of Indianapolis, Indiana, thought they had prepared their two children well in the event of an emergency. They had taught eight-year-old Deserée and her eleven-year-old brother, Danny, how to call 9-1-1. The children had strict rules against playing outside or having visitors over when they were left alone. But the couple could never have imagined the sort of tragedy that would occur on August 2, 1989.

That afternoon, Deserée and Danny were home alone watching television while their parents were at work. Georgia called around three o'clock to check on them and they told her everything was fine. A little later, three kids from the neighborhood knocked at the door—a friend of Danny's, and a boyfriend and girlfriend of Deserée's. The Troy children knew they were not allowed to have visitors, but they let their friends in, and everyone watched television.

Danny and his friend wandered into Danny's parents' bedroom. The friend spotted a rifle leaning against the wall and picked it up. Danny told his friend to leave it alone, that it belonged to his father. Danny Jr. kept the rifle accessible in case he ever needed to protect himself or his family. The rifle had been standing in the corner for so long that it had become part of the scenery.

The friend ignored Danny. He cocked the rifle twice, ejecting two bullets. Then the boy found the chamber and loaded the bullets. Deserée, her girlfriend, and the little boy came in to see what the older kids were doing and climbed onto her parents' bed. Again, Danny told the boy to put the gun down.

"C'mon, cock it twice," the boy said to Danny, "then we'll put it down."

"Stop it, Danny," said Deserée. "Put the gun down." She didn't like to see the boys playing with the rifle. But the little boy was enthralled.

"Put it down," Deserée told him again.

To appease his friend, Danny cocked the rifle twice and loaded the bullets into the chamber. He set down the rifle, then he and his friend left the room.

Now the little boy wanted a turn with the rifle. He picked it up and tried to cock it like the older boys had done. Danny ran back into the room and yelled at him to put the rifle down. But before the boy could, he accidently fired it.

Danny looked around frantically to see what the bullet had struck. He was sure it had hit a wall. Then he noticed Deserée. She was lying on the bed, bleeding. She had been shot.

While the other terrified children ran out of the house, Danny dialed 9-1-1. His call was answered by 9–1–1 operator Denisa Stevenson. Stevenson recalls the panic in Danny's voice.

"My sister's been shot! I think she's dying!" he said.

While police and fire units were dispatched to the scene, Stevenson tried to calm Danny and keep him on the line. Danny was frightened and wanted to call his mother at work, but Stevenson instructed him not to hang up.

"What's your sister doing?" Stevenson asked.

Dereeé had climbed off the bed. Danny told Stevenson that she was trying to walk to him. The dispatcher instructed Danny to make Dereeé lay down and be still. The phone reached to Dereeé, so Danny put the receiver to her ear and Stevenson instructed her to lay still and reassured her that help was on the way.

Police officers Ron Burgess, Jr. and Greg Weber arrived within minutes and knew the situation was critical when they saw Dereeé lying in a pool of blood. Paramedic Supervisor Norman Hockley arrived moments later and examined the little girl. He was astonished that she was still alive, considering her extensive blood loss.

Dereeé was rushed to Wishard Hospital, where a pediatric surgeon performed an emergency surgery. The most critical part of her wound was where the bullet had entered the right side of her abdomen and exited the left side of her chest. She had sustained extensive liver damage and a ruptured spleen. The doctor told the Troys that Dereeé had a 5 percent chance of surviving.

When Danny was allowed to see his sister in the hospital, the first thing he said to her was, "I love you."

"I love you, too," gestured Dereeé.

By all odds, Dereeé should never have made it. But she made a miraculous comeback, and less than two months after the accident, she entered the third grade, a normal, healthy girl.

"It's hard to talk about this sometimes," says Dereseé, "because it makes me remember how many times it's happened to other kids and they've died and I survived. I thought I was going to die, but I kept on pushing and pushing for myself."

Dereseé is most appreciative of her brother.

"When they told me Danny saved my life, I wanted to jump up and run to him and hug him," she says. "If he didn't call 9-1-1, I would have died on the floor."

Since the accident, Danny Jr. and Georgia's attitude about guns have changed. Danny Jr. used to think that having a gun protected his family. Now he realizes it almost cost him his daughter's life.

Pesky Python

For years, Ruth and David Spalding of Fort Lauderdale, Florida, have enjoyed watching the wildlife in the 180-acre state park that borders the backyard to their house. They never imagined that one would become a house guest.

On the night of August 17, 1989, the couple heard a strange noise in the backyard and David went outside with his flashlight to investigate. He glimpsed an enormous snake wrapped five times around a raccoon. David called to Ruth and she came outside in time to see a tail slither under their house.

David warned his neighbors, fearful for their pets and young children, who were smaller than the raccoon the snake had killed. For weeks, David sought help from local and state agencies, including 9-1-1, but was repeatedly told they didn't handle snake problems. David built an elaborate trap to keep the snake under his house, but one morning the snake broke out.

Finally, David was put in touch with Todd Hardwick, owner of Pesky Critters Relocation, a company that resolves wild animal problems in urban areas by capturing animals alive. Todd and snake expert Tom McLellan went to the Spaldings' to check out their problem. Tom took one look at the shedded snakeskin and identified the reptile as a twenty-foot reticulated python.

"What are reticulated pythons like?" asked Todd.

"Uh, they're usually nasty," replied Tom.

Tom and Todd were excited and nervous as they peered under the house and saw tunnels carved in the dirt from the python's movements. They examined blueprints and decided to crawl underneath the house to find the snake's exact location. David felt responsible for the men and thought it unwise for them to go under the house, considering the size of the snake. But Todd and Tom belly-crawled down a hole and followed the tunnels through the many chambers of the foundation until they reached the one that appeared to house the snake. The strong smell of snake musk told Tom the python was presently under the house. The men weren't foolish enough to confront a twenty-foot python alone, however, and decided to come back later with reinforcements.

The next morning, Todd and Tom returned with snake hunters Felix Velez and Joe Wasilewski and their gear. A crowd of camera-toting spectators formed as the men prepared for their hunt. Tom wasn't too thrilled about crawling under the house.

"That was the worst part of the day—getting in that little hole," recalls Tom. "Being in closed places drives me crazy. I felt this whole house collapsing in on me."

"Of course I was afraid," says Felix. "I'd be lying if I said I wasn't. But fear is part of it. We like adrenaline, we like excitement. It's something different."

As the four men crawled toward the snake's home, they blocked holes to the other chambers with large rocks so the snake couldn't escape. At the entrance to the snake's chamber, Tom peered inside and got a good look at the python. He was amazed at its size, but Todd was terrified it might strike Tom's face or throat.

The men were too far from the entrance they'd come in by to wrestle the python into a sleeping bag and drag it out. Joe, anxious to get out to open space, crawled back out with Todd to look for an alternative entrance, leaving Tom and Felix to watch the snake. Todd, extremely nervous about leaving his buddies with the snake, quickly dug a hole on the other side of the house into the python's chamber.

Todd held a long stick ending in a roped noose and was lowered under the house headfirst, while two men held on to his ankles outside the house, prepared to pull him out. While Tom and Felix prodded the python forward, Todd slipped the noose around the snake's neck. Neighbors screamed and beat a hasty retreat as Todd emerged with a twenty-foot, 250-pound python.

"It boggles our imagination," says David, "to think that for fifteen to eighteen years we were not aware the snake was here at all."

"If I had one word to sum up the whole thing," says Tom, "I'd say, 'great,' because I wouldn't have missed this for the world."

Nor would have Todd, who named his new python roommate Big Mama.

"Some of the fellas said they never had so much fun in a long time," says David. "When you're into snakes I guess that's really the greatest thing—to be able to go under a house and get one."

Rescuers Rescued

On October 1, 1989, rainy weather kept Mid-Georgia Ambulance Service busy responding to car accidents in Macon, Georgia. That afternoon, Mid-Georgia received the report of a multi-car accident. Paramedic Steve Parker and his partner, Tim Powell, hopped into their ambulance. In the event of multiple injuries on the scene, paramedic Neil Stevens and his partner suggested they also respond as backup.

The two ambulances kept in radio contact as Stevens followed Parker in the pouring rain. The roads were slick, and at one point, Parker hit a dip and went into a skid. He regained control of his vehicle, then radioed to Stevens to warn him of the upcoming danger zone. Stevens slowed to avoid a problem, and the two ambulances continued on.

When Parker went around a curve and lost sight of Stevens, he became a little concerned, so he radioed the dispatcher to tell her what had happened. Parker wanted to turn around and drive back to find his co-workers, but training had taught him that upon being dispatched to a call, he must respond or possibly be held liable for patient abandonment.

Stevens had not rounded the curve because as he drove into it, the wet streets caused his ambulance to skid off the road. The vehicle struck a fire hydrant, rolled over three times, and landed on its side. The fire hydrant ripped a hole in the gas tank and the ambulance erupted into flames. Neither Stevens nor his partner were seriously hurt, but they were trapped inside the vehicle and faced burning to death unless they escaped immediately. As flames spread around the ambulance, Stevens desperately struggled to release his seat belt while his partner tried to kick out the windshield.

Norris Thomas, a Georgia Power Company trouble-shooter, was on his way to work when he saw the fiery wreckage. He jumped out of his pickup and ran over to the vehicle, where he saw a crack in the windshield just wide enough to stick his fingers through. Thomas broke the glass, tore the windshield from the car, and reached inside. He grabbed Stevens's partner, who was twice his own weight, pulled him out, and threw him away from the vehicle.

"I said to myself, there's no way I'm going back in that," recalls Thomas. "And by that time, I was back in there."

As flames consumed the front of the ambulance, Thomas reached inside, somehow managed to release Stevens's seat belt, and yanked him to safety.

"If we hadn't gotten out in the next ten seconds," recalls Stevens, "we'd be six feet under. We owe our lives to that man."

As firefighters arrived on the scene to extinguish the blaze, the paramedics looked for Thomas to offer their thanks. A bystander told them he had left. Thomas, concerned he'd be late for work, had quietly walked back to his truck and driven off without giving his name to anyone.

"But you don't understand," said Stevens to the bystander, "I've got to go thank this man. He just saved my life."

By the next day, everyone in Macon was trying to find the mystery hero. A co-worker of Norris Thomas's heard a report over the radio that the paramedics were looking for their rescuer. The co-worker had already learned that Thomas was their man. He also knew his buddy was the reserved type, but he wanted him to get the credit he deserved, so he asked Thomas for permission to give his name to Mid-Georgia Ambulance. Thomas consented.

Stevens and his partner finally got to personally thank Thomas for saving their lives. When they met him, they were shocked that a man of such average size had mustered the power to pull them out.

"I never had that feeling before," recalls Thomas. "It's just like a force, like energy. And I know if it weren't for that instinct, it wouldn't have happened."

Thomas asked the paramedics to play down the whole episode.

"But you did a great thing," Stevens told Thomas. "My job is to go out and save people's lives every chance I get. That's what you did for me."

"I was stunned," says Thomas. "I never did realize that. I never did think about it that way."

Remorseful Burglar

At one-thirty on the morning of May 19, 1989, Los Angeles Police Dispatcher Pam Graves answered a call for help unlike any other police call she had ever heard.

"I broke into a glass shop and I can't get out," said the caller to Graves. "There's too much glass in my way in the doorway. I know I shouldn't be here. I'll turn myself in. If I step over this glass it'll cut me very seriously."

Yeah, right, thought Graves, a burglar calling to turn himself in. She was fairly sure the call was a prank, but she took the man's address and told him she'd send someone over.

"It won't take long, will it?" he asked Graves nervously.

"I hope not," she responded.

The unlucky caller had in fact broken into a glass company. He had smashed the glass door with a shovel, climbed inside, and looked around for something to steal. Finding nothing, he went out the back door, only to encounter two snarling rottweilers guarding the company's property. He ran back inside and slammed the back door shut. But he was afraid to crawl out the way he'd come in because he'd broken a two-foot-diameter hole in the glass door, which was now a ring of sharp, deadly shards. Feeling trapped, the burglar called 9-1-1 for help. He told the dispatcher he was unarmed.

When Officer Bob Barnes and his partner first got the call, they looked at each other and said it must be a mistake. As they headed to the scene, they wondered if it could be a setup or an ambush situation, even though it did appear that the suspect had made the call.

Sergeant Mark Mooring also responded to the call with his partner, police dog Sergeant Joe Friday.

"The danger in this call is that it didn't make any sense," recalls Sergeant Mooring. "Suspects don't call you and tell you where they are. He might be a psycho, and if that's the case, we're gonna need some extra people."

Officer Barnes and his partner were the first to arrive on the scene, followed by a police helicopter that circled to spotlight the area. Sergeant Mooring and the K-9 unit also arrived, as well as four additional backup units. It was tactically a bad situation for the officers because they couldn't see anything through the windows. They had no idea what to expect.

Officers surrounded the property, then approached the building. Sergeant Joe Friday, the police dog, was sent up to the hole in the glass door. He immediately smelled the suspect and began to bark and growl loudly.

"Keep the dog away from me! Keep the dog away!" yelled the suspect from inside, evidentally quite frightened of dogs.

The burglar surrendered to police and thanked them for getting him out of the building.

"This was definitely one of the easiest felony arrests I've ever had," says Mooring. "It doesn't take a lot of brains to be a burglar, but this guy had none."

The suspect pleaded guilty to vandalism and trespassing and was sentenced to two years probation and 150 hours of community service.

"The suspect had to be rescued from a situation he created himself—from a crime," laughs Mooring. "I like that. That's a good ending."

Fernie's Heart Transplant

On July 24, 1982, Yvonne Ayala of Los Angeles County gave birth to a baby boy, Fernando "Fernie" Ramos. When Fernie was six months old, he was diagnosed with a rare condition that causes the heart muscle to deteriorate and results in death. The only thing that could save his life was a heart transplant, but doctors were unwilling to perform the experimental surgery. They told Yvonne to make Fernie comfortable at home because he would die within six months to a year. But Yvonne had no intention of giving up on her son, and she continued to search for a miracle.

Fernie was put on medication and defied his doctors' prediction—he celebrated his second birthday. That same year, Yvonne saw a news report on television about a ground-breaking heart transplant performed on an infant by Dr. Leonard Bailey. She thought her prayers had been answered.

Yvonne got Fernie an appointment with Dr. Bailey's heart transplant team at Loma Linda University Medical Center in Southern California. Dr. Bailey told Yvonne that Fernie could indeed be a candidate for a heart transplant—but he would have to be within six months of death before he could be placed on a nationwide transplant list.

"So I waited," says Yvonne. "I thought, well, the hard part was over. But it wasn't."

For the next three years, Fernie's condition remained stable, but in January 1988, he began to deteriorate rapidly. Over four months, he was hospitalized three times for pneumonia and congestive heart failure. Now almost six years old and weighing only thirty-three pounds, Fernie was finally placed on the transplant list and transferred to Loma Linda. Doctors thought that if a heart was not found within a few weeks, Fernie would die. Yvonne turned to the media and made a desperate appeal to the public to help save her son.

The next day, less than twenty miles away, Audrey Radcliffe and her family received tragic news of a car accident. Audrey's brother, Dave, had been killed, and Dave's nine-year-old son, David, lay in the hospital, brain–dead. Doctors asked if the family would consider donating little David's organs.

"I loved my son so much," says Dave's father, Lee Denehy, "and little David. And we thought that rather than it be just a waste, there's a lot of other people waiting for somebody's heart and they're going through misery. The type of person Dave was, always helping everybody, this is what he'd want."

Fernie's long wait had finally ended. David's heart would be his last hope. As Fernie was wheeled into the operating room, Yvonne said good-bye, thinking this would probably be the last time she would ever see her son. She also thought about the donor family's sorrow and wondered if they knew where the heart was going.

Fernie's transplant operation was a success. David's heart was a perfect fit and it started beating immediately upon implantation.

Says Dr. Bailey, "Every time we do a heart transplantation, it's an amazing miracle. You actually lift a heart out of someone and put it to sleep temporarily, and then implant it in someone you know has all the hopes and dreams for a future. And to have that heart come alive again and do what it's supposed to do is a genuine miracle. I'm awed by it."

One week later, Audrey and her family were eating dinner when they saw Yvonne on television, telling reporters that Fernie was alive thanks to his new heart. She had a special message for the donor's family.

"I know you're watching me, and thank you," Yvonne said, looking directly into the camera.

Right then and there, Audrey knew where David's heart had gone. Although hospital policy prohibits the release of the identities of transplant donors and recipients, Audrey located Fernie and Yvonne. The two families became very close, and today a photograph of David sits on Fernie's nightstand in a heart-shaped frame.

"Fernie's like one of the family," says Audrey. "He's not my nephew, but he's a little boy who got a chance to go on. I can't thank Yvonne enough for letting us be a part of his life."

Dr. Bailey hopes that heart transplant recipients like Fernie will convince more people that organ donation is a good cause.

"We can make Fernies happen everyday if we really work at it in this country and around the globe," says Dr. Bailey. "We can reverse some of the tragedies that happen and make them all victories."

RN's Rough River Ride

Nurse Kris Walker and five girlfriends from Valdez, Alaska, were excited as they suited up in rain gear, boots, and life vests. It was July 3, 1990, and the six nurses were preparing for what was supposed to be an hour of fun; white water rafting down the turbulent, rocky, freezing-cold Lowe River.

The rafting guide had warned the women that the hard part of the ride would be to avoid hitting the pillars of the bridges they would pass under. As the group set out, their excitement built as they quickly encountered rough water. The raft was moving fast as it rounded a bend and approached a bridge.

"Dig in!" shouted the guide, urging the women to paddle hard.

But the current swept them straight toward a pillar. The raft struck it broadside and flipped over. The group was thrown into the 38-degree water and carried away in the rapids. Kris held on to the raft until the current pulled her under.

"The water was just sucking me down," she recalls, "and then it would spit me up just enough to get a breath of air. The river was totally in control."

Suddenly, Kris found herself in a calm spot and climbed onto the rocky shore. The guide and two others were also swept into shallow areas at the river's edge. Exhausted, they pulled themselves onto the bank. But Mary Lee Hayes, Joan Tate, and Kristin Ellingson were still lost in the rapids.

Mike Buck, an expert kayaker, and owner of the rafting company, happened to be driving along the river on his way out of town when he saw a scary sight—three rafters floating downstream without their raft.

"It was just unbelievable to me," recalls Mike. "We'd had only one flip in nine years."

The guide had grabbed a rope from the raft and stood on shore. He threw the line to Kristin as she floated past, but she missed. Mike yelled to the guide to drive downstream with him. He had to rescue the women quickly, before they succumbed to hypothermia. In water that icy, it would be a matter of minutes before their core temperature would drop, causing incoherence, loss of use of their limbs, unconsciousness, and ultimately, death.

Mike got his kayak from the van's roof and threw on his water gear, while the guide threw a rope to Kristin, who was approaching. This time, Kristin grabbed the rope and the guide reeled her to shore. Mike set his kayak on the water and jumped into it. There was no time to put on a helmet or life vest because Joan and Mary Lee had already floated past. Mike paddled furiously through swirling rapids to catch up to them.

Mike reached Joan first. She wrapped her arms around the front of the kayak and hung on while Mike paddled to the bank. Joan, exhausted and mildly hypothermic, climbed the bank and joined her friends in Mike's van on the side of the road.

"The bottom dropped out of me," recalls Kristin, "when I didn't see Mary Lee's face. She's a real dear person to me, and I felt like she might be dead."

Mike scanned the river for Mary Lee and soon saw her floating in the distance, her head barely above water. He raced to close the gap before she lost consciousness. When he reached her, Mary Lee flung her arms around the kayak and managed to hang on until they reached the bank.

In the van, the women warmed Mary Lee, whose body temperature had dropped well below normal. She was confused and unable to see.

"She looked terrible," says Kristin, "but she was alive! And I knew we were going to have one hell of a story to tell."

Mary Lee was treated for hypothermia at a local hospital and released the same day. Her eyesight gradually returned, and she suffered no permanent ill-effects as a result of having spent fifteen minutes in the icy water. The women, especially Mary Lee, feel incredibly lucky that Mike spotted them that day.

"I wasn't thinking there was a way out of it," recalls Mary Lee. "At that point, I saw an angel come. He was very bright and had blond hair and a big smile. I remember the smile. Mike Buck is a very special person to me. He saved my life."

"It was really a miracle," says Mike, "that I was there at the right place with all the right equipment. The angels were definitely with us that day. There's no question about it."

High Voltage Save

On the morning of April 14, 1992, thirty-nine-year-old Herbert Frederick of Murray, Kentucky, was rushing to get to class on time. Herb was less than three weeks away from achieving his dream of graduating college, but this morning he was running late. Being a father, holding down a full-time job, and going to college made it tough to stay on schedule each day.

Herb was driving on a busy two-lane highway, stuck behind four other cars as he rushed to the nearby university. He thought he saw his chance to pass the cars ahead of him. He pulled into the oncoming lane, but heading directly toward him was a pickup truck traveling in the opposite direction. Driver Steve Ernstberger saw Herb coming and tried to swerve off the road. Herb narrowly missed hitting Ernstberger, but he lost control. His car sailed off the road, became airborne, clipped a utility pole, and crashed into a tree.

Ted Potts heard the crash from inside his house and called 9-1-1. Then he ran outside to help.

"That old boy doesn't have a chance," Potts thought to himself when he saw the mess Herb was in.

The utility pole had fallen to the ground, and its power lines sagged across the length of Herb's car, hanging precariously three to four feet above the car and the road. Nearby, the lines had become tangled in a tree. With each power surge, sparks flew, the tree limbs burned, and the lines dropped a little farther toward the roof of the car. Ernstberger and other motorists stopped to help, but fear of electrocution kept everyone away from Herb, who was slumped unconscious at the wheel.

Rescue units from the Murray Fire Department, led by firefighter/EMT Brad Haugh, arrived on the scene, followed by Murray-Calloway County Hospital Ambulance Service with paramedic Owen Moore and EMT Kenny Collins. Haugh quickly learned that the firefighters had a bigger problem than anticipated. Several yards away, a trailer house had caught fire as a result of the downed power lines. Haugh told his men not to go near the trailer if there was a risk of electrocution.

Haugh wondered how he and his men were going to rescue Herb without risking their own lives. His training had taught him never to approach an area with live power lines.

"You just don't run into a scene where there's power lines involved," says Haugh. "You just do not do it. Never."

Paramedic Moore was also anxious about approaching Herb's car, but he felt he had no choice.

"Well, we can't just let him die," he said to Haugh.

Moore went to his truck to get some equipment. That's when he noticed a fire burning under Herb's car. If rescuers were going to act, it would have to be this second. Moore, Collins, and Haugh hurried to the car and pulled Herb out as best they could, knowing that if they moved him the wrong way, they could paralyze him or kill him. But speed was of the essence to prevent all of them from being electrocuted. The men slipped a C-collar around Herb's neck and slid him onto a backboard to protect his spine. By the time they reached the ambulance, the power lines had fallen, and the car was completely engulfed in flames.

Herb was admitted to Lourdes Hospital where he underwent reconstructive surgery to repair his shattered right leg. He also began the road to recovery from a severe head injury that caused permanent loss of sight in one eye. His wife, Paula, was preparing herself for her husband's uncertain future, but Herb made a remarkable recovery. Three weeks after his accident, he was transferred to a rehabilitation center.

"This is an amazing recovery," says Herb's neurologist, Dr. John Colby. "Herb's a lucky guy. He's coming back further than anybody ever thought he would. Had his rescuers waited until everything was just right, Herb might not be around right now."

Dr. Colby attributes Herb's speedy recovery to his determination and positive attitude, and to the support he received from Paula and his family.

Paula also learned a lesson the hard way.

"After the accident, I just wanted to follow people and say, life's too short," says Paula. "You don't need to get there so fast. Come look at my husband. See what happened to him."

One year later, Herb received his college degree.

"I guess I'm a cat with nine lives," he says.

Mallard Madness

On June 17, 1990, numerous visitors to Calgary, Canada's Prince's Island Park noticed a mallard duck with a plastic six-pack ring entangled around its head and neck. Park-goers, accustomed to feeding the mallards, could tell that the duck was having trouble eating. Over the next couple days, visitors tried to lure the duck with bread crumbs in the hope of catching it and removing the plastic ring. The news media picked up the story and the duck was named—Ed.

Department of Fish and Wildlife officers made several failed attempts to capture Ed. Officer Jack Morrison, involved in the rescue effort, received numerous phone calls from concerned citizens who were worried that Ed would starve or strangle to death.

"One fellow suggested we should take some grain and soak it with whiskey and get the duck intoxicated," recalls Morrison. "Another said that we should dive beneath the water. I believe somebody tried to scuba dive and grab the duck from underneath."

As word of Ed's predicament spread, Colleen Ferguson, a local television news reporter, was sent to cover the story.

"We were almost held hostage by that duck," says Ferguson. "This story really excited people. Everybody wanted that duck caught."

When all attempts to rescue Ed failed, the Department of Fish and Wildlife called in an expert. Jeff Marley was the manufacturer of a "net launcher," a gun device that shoots a weighted net. It was used to capture animals when tranquilizer darts could not be used. Marley did not want to net Ed on the water because there was a risk he might drown, so his first plan was to capture him on land. But Ed got smarter as the week wore on and avoided being lured onto the banks.

Finally, Marley took a shot at Ed while he was on the bank, but Ed escaped just as Marley got his hands on him. Another day, Marley decided to take his chances and shoot Ed while he was in the water.

"We caught a duck," recalls Morrison, "but not the right one."

Capturing Ed proved harder than anyone had imagined. Ed eluded his captors for several days by staying out of sight. Meanwhile, as his fame grew and more crowds gathered in the park, Ferguson continued to follow her story, and would call Marley whenever she spotted Ed.

"I knew I'd catch him," says Marley. "It was just getting the right shot."

On July 4th, Marley and his assistant set up an ambush, pushing Ed upriver with light harassment. Ed fell for it and ended up right in front of Marley. Marley took aim, but he missed his shot.

A little later, Marley spotted Ed in the water near a pedestrian bridge that spans the river. The bridge was filled with onlookers and Ed was wary of getting near anyone. It looked as if he wanted to swim under the bridge, so Marley decided to surprise Ed with a shot from above. He went out on the bridge and waited until Ed passed underneath and emerged on the other side. Then Marley fired. This time, he netted Ed.

Concerned about Ed being dragged under water by the net, Marley didn't waste time running down to the bank. He jumped off the bridge, thinking the water level was still five feet and intending to land on his rear end. Unfortunately, the water only came up to his knees and Marley landed on his left leg. He carried Ed out of the water and Ferguson cut the six-pack ring off Ed's neck. Ed then flew to freedom as the crowd cheered.

There was only one casualty that day. Jeff Marley broke his leg jumping off the bridge and was in a cast for a month. Two weeks after he rescued Ed, Marley received the City of Calgary's first Environmental Awareness Award. Marley hopes that in the future, six-pack holders will be made biodegradable so something like this won't happen again.

"A lot of people wondered, why the big effort for one duck?" says Ferguson. "Ed was special because it was man that put him in that position, and he couldn't get himself out of his predicament. We pollute too much and this is the kind of thing that can happen."

Russia Rescue

In a remote part of Siberia lies the seaport of Magadan, home to the Avdeyenko family. On the afternoon of September 17, 1990, eight-year-old Anton Avdeyenko and his best friend, Maxim, decided to make a fire to toast bread and potatoes. Anton found some matches he'd hidden in his apartment building, then he and Maxim headed next door to a construction site to gather wood. There they found large barrels of liquid that smelled like oil or shellac.

"This liquid will burn really well," said Maxim, as they tipped a barrel so Anton could fill a small can.

The fluid poured out hard and fast—onto Anton's clothes. Maxim told him to go change, but Anton was afraid his mother would get mad at him.

The children were sitting around their fire when a piece of burning paper flew up and landed on Anton's leg. Instantly, he erupted into flames. Anton started running wildly. Anna Avdeyenko was in her kitchen when she heard screaming outside and rushed to her balcony.

"I saw a burning torch," recalls Anna, "but I couldn't see a face because of the fire. Then I saw a red cap and figured out it was Anton."

The driver of a passing water truck stopped and poured water on Anton to extinguish the fire. In tears, Anna carried her son upstairs and cut off his clothes, but his polyester pants had melted onto his skin. He was a horrifying sight.

"It's impossible to say how a mother feels when she's losing her child right in front of her eyes," says Anna. "My heart broke into little pieces and couldn't be put back together again."

A passing truck drove Anton and his devastated parents to the hospital.

"I wish I could have put myself in his place," says Anton's father, Vladimir, "and do the suffering for him."

Doctors at the Magadan Regional Hospital rushed Anton into the first-aid room where they removed dead tissue, disinfected his wounds, and eased his pain. Since most burn victims die of infection, Anton was isolated, even from his parents. Despite doctors' best efforts, Anton deteriorated. He had received burns over 33 percent of his body—it was crucial that he be treated at a burn unit. But there was no vehicle in which the critically injured Anton could safely ride four thousand miles to the nearest burn unit.

Vladimir was determined to find help. He approached Larry Rockhill, an American foreign exchange teacher in Magadan. Rockhill contacted Dr. Ted Mala, Director of the Institute for Circumpolar Health Studies, in Anchorage, Alaska. Dr. Mala agreed to try to cut through red tape for official government permission for Anton to be treated in the United States. If he succeeded, it would be the first time anyone was permitted to leave what was then the Soviet Union to receive medical treatment.

Another contact in Juneau, Alaska, Betty Johnson, made hundreds of phone calls over a three-day period to secure the Shriners Burn Institute in Galveston, Texas, to provide treatment: Rocky Mountain Helicopters to provide transportation: and Anchorage's Providence Hospital medivac team to provide in-flight medical care—all for free. Dr. Mala received the U.S. and U.S.S.R governments' permission for Anna and Anton's journey.

Flight nurse Marilyn Belanger says of the historic trip, "When we were flying over, we didn't think of Americans and Russians. All we thought of was a patient and a life."

Surgeon Paul Waymack of the Shriners Burn Institute, recalls that Anton had one of the deepest burns he'd ever seen. In Russia, the wounds had become extensively infected with bacteria, and now reached down to muscle and bone. Anton was administered massive doses of antibiotics, and, because so much of his own skin was burned, wounds were surgically treated with cadaver skin.

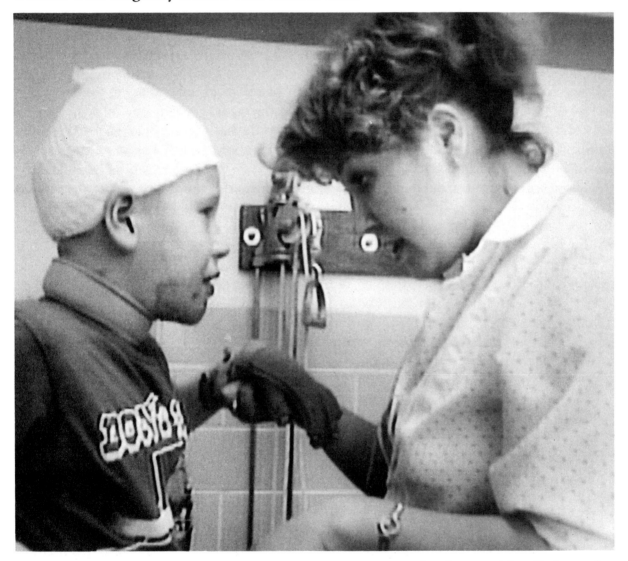

Anton began to improve, and it soon became apparent that he would pull through.

"The happiest moment," recalls Anton, "was that I got back up on my feet, and that I came out alive. The doctors who treated me in Texas were the best of all."

After three weeks in the United States, Anton returned home to Magadan and to a joyous family reunion.

"Now I feel very well," says Anton. "I want to thank all the people who helped me fly to America. And I learned that matches are a bad joke. I don't think I will be playing with fire again."

Runaway Truck

Truckers Wayne Waite and Louis Gallant had been best friends for years. On January 14, 1984, their friendship was tested to the limits in the mountains high above Nova Scotia, Canada.

Having spent the night at a truck stop, Louis stepped out of his cab's sleeping compartment the next morning and spotted Wayne's rig. Louis woke up his buddy and they had breakfast. When the men discovered they were both headed in the same direction, they decided to drive in tandem so they could chat over their CB radios.

It was clear and sunny as the men headed out of the truck stop at six in the morning. Wayne led the way in his eighteen-wheel refrigeration truck, which was packed with 23 tons of frozen food. Louis followed in his flatbed semi that was hauling 17 tons of steel. Although they both drove for the same company, it would be the first time the friends would be driving together.

As Louis and Wayne climbed the steep mountain road, Louis told Wayne on the CB that he wanted to pass. This was highly unusual, because Louis never liked to be in the lead. But for some odd reason, this morning he didn't feel like following. Wayne signaled to him when it was safe, and Louis accelerated and passed.

The truckers approached the top of the mountain and decelerated, preparing for the treacherous descent in which they would maneuver numerous curves and 180-degree turns. They were driving along at twenty-five miles per hour for less than a mile, when Wayne applied his brakes. He was stunned.

"I got no brakes, buddy," Wayne radioed Louis.

"You're joking," Louis responded.

"I ain't kidding. I got no brakes."

Wayne pumped his brakes, but they still wouldn't work.

"I'm going to try to dynamite it," Wayne told Louis.

Wayne "dynamited" it—applied his emergency brake—but it wasn't working either. He was picking up speed, and he didn't know what to do. A truck as heavy as this one couldn't possibly make it safely to the bottom of the windy mountain road without any brakes.

"Run her into the back of mine," radioed Louis. "I'll try to get you stopped."

Less than one mile away was a straightaway that stretched for a half-mile. If Louis were going to stop Wayne's runaway truck, it would have to be here, because coming up was a horseshoe curve that Wayne would never be able to maneuver without slowing down.

But before reaching the straightaway, Wayne had to steer through a series of S-curves without braking. As Wayne's truck rapidly gained speed, Louis was forced to accelerate to stay in front. The men barreled through the curves, traveling close to forty miles per hour instead of the safe twenty-five.

"I'm going too fast," Wayne radioed. "I'm going to have to jump."

"No," responded Louis. "You'll kill yourself."

The truckers approached the straightaway. Louis slowed to let Wayne catch up to him. As both men braced themselves, Wayne aimed for Louis's truck. Wayne rammed into the rear of Louis's semi, now going about sixty miles per hour, and Louis slammed on his brakes. The two trucks slid for half a mile, tires squealing and rubber burning, then screeched to a halt less than five hundred feet from the horseshoe curve. The men stepped out of their cabs, shaken, and hugged each other.

Wayne suffered only a few bruised ribs and Louis escaped unharmed.

"I thought it was all over," recalls Wayne. "I thought this was my last trip. At that mountain, if your brakes hold one vehicle you're thankful. But to hold two—it's a miracle."

"I knew I had one chance to save him," says Louis. "I just wanted to do it right. If he had hit at an angle or lost control, it would have probably taken me, too."

Louis and Wayne, best friends before the incident, have become even closer.

"Louis would do it for anybody," says Wayne. "He's that type of person. He was a friend before this, but he's a special friend now."

Wayne still drives trucks for a living. But three years after that fateful day, Louis retired from trucking.

Logger's Baby

Ken Leppert and Ken Murdock of Klamath Falls, Oregon, friends since high school and fellow mechanics for a logging company, were on the job the morning of December 7, 1989. The temperature was below freezing and mist hung in the air as the two Kens drove down a remote logging road toward town. They reached an open gate, and Ken Leppert jumped out of his truck to close it. As he locked it, he noticed a curious bundle lying on the forest floor. Leppert went over to investigate and saw a folded blanket. He started to unwrap it and saw blood. He ran back to the truck.

"Come and tell me this isn't what I think it is," he said to Murdock.

The men approached the blanket and Murdock started to unwrap it. The first thing he saw was a placenta and umbilical cord. Murdock continued to unwrap the blanket, and neither man was prepared for what was inside. They had found a newborn baby boy. The men hurried back to the truck with the infant, whom they thought was dead. They radioed their office to call the sheriff or state police. While on the radio, the two Kens heard a whine. They checked the baby and he moved one eyelid. He was alive—barely. The men requested an ambulance.

The blanket was soaking wet and the men were scared that the baby, who was rigid and blue in color, was on the verge of freezing to death. Leppert peeled off his sweatshirt and wrapped it around the baby for warmth. Then they carried him into the truck and blasted the heater. Deciding they couldn't afford to wait for the ambulance, they drove down the road to meet it. As they rode, the two Kens talked to the baby and told each other, and him, he was going to be okay, but in the back of their minds, they both felt he was too far gone.

Sheriff's Deputy Tom Johnson was heading to the scene when he encountered their truck.

"It really hit home hard," recalls Deputy Johnson, "because here's a brand-new baby and it appears the woman had the baby possibly right there and then just wrapped him up and just discarded him like so much baggage."

Moments later, the ambulance arrived. The Kens had named the baby boy Benjamin, and Ken Murdock gathered him in his arms and went around to the back of the ambulance.

"They had to chase him out of the ambulance," says Leppert. "He was going to go with him."

"You form a bond real quickly with a baby like that," says Murdock. "You sit there and hold him so tight. I didn't want to put him down. It was a funny feeling, you know, like you feel he's yours. It's not something you want to just walk away from. It's something you want to take home with you and hold tight."

When baby Benjamin arrived at the hospital, his heart had stopped and his core body temperature was so low it didn't register on the thermometer. Pediatrician Dr. Charles LaBuwi arrived from his office, but he had told his staff he would be returning shortly because he didn't think Benjamin would survive.

"I thought the dear Lord would determine whether this baby was going to survive or not," says Dr. LaBuwi, "and I'd just do my part."

Amazingly, baby Benjamin regained a heartbeat and his condition started to improve. Two hours after his arrival at the hospital, his body temperature had elevated to ninety-four degrees, he was crying, and was well enough to be transferred to the nursery.

Ken Leppert called the hospital to see how Benjamin was doing.

Recalls Ken, "They said, 'Well, you can come and see this baby twenty-four hours a day if you want to. We consider you guys family.' We got instantly attached to him. We both just fell in love with him."

While Benjamin was recovering, hundreds of calls poured in from couples wanting to adopt him. After two weeks in the hospital, Benjamin was adopted. Because Ken and Ken had found him so close to Christmas time, they were asked by reporters if there was any religious significance.

"Ken and I looked at each other," says Leppert, "and Ken said, 'No, but it does make you believe in Santa Claus.'"

Lansing Stabbing

On October 16, 1988, Prima Church of Lansing, Michigan, arrived home after work, unlocked her front door, and went inside to unwind for the evening. She made herself a plate of nachos and carried it into her bedroom. Suddenly, Prima was confronted by an intruder wielding a knife. The robber blocked Prima's escape through her front door and stabbed her in the chest, knocking her on the floor. The man grabbed her purse and threw it at her. Prima pulled out the little money she had and threw it at her assailant, who then ran out of the house.

Prima picked up the phone and dialed 9-1-1.

"Hello," she said to dispatcher Paul Bouldin. "My name is Prima Church. I've just been stabbed. I'm bleeding all over the place."

Bouldin instructed Prima to get a towel and apply pressure to the wound in her chest. Prima spoke calmly at first and described the suspect to Bouldin, but as her condition deteriorated, panic set in.

"Just calm down," Bouldin told her.

"I'm trying to keep calm, sir," Prima replied. "He was here waiting for me. He drank a beer—"

"Okay, ma'am," interrupted Bouldin, anxious to dispatch rescuers to her aid, "just hang on here, just a minute."

"I'm bleeding!"

"Yes, ma'am, I'm going to get someone en route."

Bouldin dispatched police and paramedics to the scene, then stayed on the line with Prima, trying to comfort her and help her stay calm until help arrived. Prima told Bouldin that the front door was locked.

"Just stay where you are," he responded. "Don't move."

"Oh, I hurt," she moaned.

"I know you're hurting," said Bouldin. "Nice and easy, breathe in through the nose and out through the mouth."

"I'm going to pass out," replied Prima, who lay on the floor, "I can feel it right now."

Bouldin could hear Prima's voice quiver and grow weaker, and he was afraid she was going to die while he had her on the phone. But he knew he had to stay calm because his voice was her only link to outside help; if he didn't stay calm, she wouldn't either.

"Just concentrate on your breathing," he repeated.

"It hurts."

"Go ahead, if you feel like crying, just let it out. Sometimes crying will help."

Police Sergeant Nancy Small was the first to arrive on the scene and found the front door locked. Bouldin told Prima that police officers had arrived and might break a window to get inside. Sergeant Small went around the back of the house and saw an open window, which she presumed the assailant had used to gain entry. She climbed in through the window and hurried to open the back door for fellow officers and then the front door for paramedics.

Paramedics found Prima bleeding profusely from a chest wound caused by a knife blade. She was immediately transported to the hospital, where that evening she received a visitor, dispatcher Paul Bouldin, who wanted to meet the voice over the phone and wish her well. Prima was treated for a punctured lung and stab wounds and released from the hospital four days later.

Had she been unable to call 9-1-1, paramedics say it's likely that Prima could have bled to death.

"As bad as the situation was," says Paul, "at least something good came out of it. She knew to dial 9-1-1 and it worked."

Prima now lives with her brother and has become extremely cautious, but she says she's gone on with her life.

"I can't live wondering and looking over my shoulder all the time," she says.

When Prima returned to work, her company asked if they could honor dispatcher Paul Bouldin. Prima was delighted, since she feels she owes her life to him. Bouldin was honored at a banquet attended by one thousand people.

"They asked me to stand up," recalls Bouldin, "and they clapped and they clapped and they clapped. If I was anywhere else, I probably would have started crying. I was overwhelmed that everything I wanted to do had made a difference. I always had the feeling that my goal in life was to help people, and with 9-1-1, I can do that. Whatever emergency you have, you can call 9-1-1 and we'll be there."

Stephanie Saves Mom

Shortly after eleven p.m. on May 26, 1991, Mary Jane Westenberg of Dearborn Heights, Michigan, was home alone in her bedroom with her daughters, five-year-old Stephanie and three-week-old Lindsay April. As Mary Jane set the baby in her bassinet, she began experiencing chest and abdominal pains. She was having trouble breathing, and the pains were getting worse. Mary Jane moaned to Stephanie to call 9-1-1.

Before Stephanie dialed the phone, she unlocked the front door and turned on the porch light so rescuers could get inside.

Stephanie's call was answered by 9-1-1 dispatcher Ken Serwatowski. At first he couldn't understand the young girl on the other end of the line because she was crying so hard.

"My mommy's having chest pains," she repeated.

Serwatowski immediately dispatched an ambulance from the Dearborn Heights Fire Department. EMTs Bill Branham and Dan Lamarand rushed to the scene.

"The worst thing I thought," recalls Branham, "was, don't let it be a heart attack. That's the hardest part of our job, to see that happen in the presence of family."

Meanwhile, Serwatowski tried to calm Stephanie and keep her on the line. When he learned that she was only five, he thought about his own young nieces and nephews, and he felt the most important thing he could do was reassure her that help was on the way. To get Stephanie's mind off her emergency, he talked to her about other subjects.

"Where did you learn how to use 9-1-1, in school?" asked Serwatowski.

"Yes, but I never used it before," she replied.

He could hear the baby crying in the background.

"Do you have a little brother or little sister?"

"Little sister."

"What's her name?"

"Lindsay April."

"You're going to have to help your mom take care of her now, you know that?"

"Yes."

"Is that your first sister?"

"Yeah."

"I have five," said Serwatowski.

Stephanie giggled.

Serwatowski heard the sirens in the background and told Stephanie she could hang up the phone.

Stephanie felt great relief when the rescuers arrived because she knew they would take good care of her mother. The EMTs were concerned about Mary Jane's complaints of chest and abdominal pains and thought she might be suffering a heart attack. As Branham attended to Mary Jane, Lamarand picked up Lindsay April, who was crying hysterically, and handed her to Stephanie. He told her she was very brave and courageous to have called 9-1-1.

Mary Jane's mother, Barbara Keller, arrived to take care of the children as the EMTs got Mary Jane into the ambulance.

Mary Jane was admitted to Garden City Hospital where tests revealed she was suffering from a gallbladder attack. She underwent surgery to remove the gallbladder, but had it not been for Stephanie, things might have turned out differently.

"I was worried that my mom was going to die," says Stephanie, "but she didn't. My hope came true. I learned if I ever have another problem, I already know what to do for me and my whole family, which is call 9-1-1."

Stephanie praises dispatcher Serwatowski.

"He seemed like a real nice man," she says. "He just talked to me a lot to make me talk better because I was crying."

"I was really amazed," recalls Serwatowski. "She handled herself better than a lot of the adult callers that we have."

On June 10, 1991, the city of Dearborn Heights presented Stephanie with an award for heroism for helping to save her mother's life.

"I don't feel like a hero," says Stephanie, "except that I just feel like I saved my mom."

Mary Jane says, "Stephanie will always be known as my little hero."

Alpine Ski Crevasse

Twenty-year-old Jurg Friedli joined five other friends on April 4, 1991, for a day of ski touring on the glaciated mountains high above Zermatt, Switzerland. Jurg, a downhill skier, had never been ski touring, so he followed the lead of his friends, Christian Brunner and Daniel Weber, who were experienced in back country skiing. It was a magnificent day as the group climbed to the summit of the 13,000-foot glacier. At the top, the men rested and enjoyed the spectacular view before beginning their descent behind a larger group led by a mountain guide.

It was important for the skiers to stay on the trail to avoid falling into crevasses—deadly, invisible cracks in the surface of the ice. But while making a turn, Jurg inadvertently veered off the trail and skied over a crevasse. He grabbed to the edges of the crack as he fell, but the snow gave way and he slipped below the surface.

Two friends saw Jurg disappear from sight and yelled to a third friend to get to the mountain guide who was leading the group ahead. The mountain guide radioed the Air Zermatt Rescue Team, which dispatched a helicopter to the scene. Daniel figured that Jurg had fallen a short distance and expected a problem-free rescue.

"But we looked into the hole, and it was so deep!" recalls Daniel. "It was a shock. We thought, maybe he's dead."

Jurg had fallen seventy-five feet below the surface between two narrow walls of ice. His friends wasted no time hauling out their ice-climbing gear, but Daniel worried that Jurg might become hypothermic.

It took a half-hour for Jens Zollhöfer, using ropes, to lower Christian down to Jurg. Jurg was still alive, but he was firmly wedged between sheets of ice in a space so narrow that he had difficulty breathing. Christian was panicked—he couldn't see any way of raising Jurg out of the crevasse.

"I had a picture in my mind," recalls Christian, "seeing Jurg's mother. And I told myself I could no longer look into her eyes if I didn't bring her son back to her."

Until the helicopter's arrival, which was slowed by strong winds, the only thing Jurg's friends could do was to chip away at the ice with a pick.

More than one hour later, the Air Zermatt Rescue Team arrived on the scene. Because Jurg's boots, skis, and hands were frozen into the ice, Alpine ski guide and rescuer Kurt Lauber realized a powerful ice air hammer was needed to chip away large chunks of ice.

By the time the air hammer was flown in, Jurg had been lodged in the crevasse for two hours. He was feeling faint and sleepy and talked about giving up.

"C'mon, Jurg! Behave yourself!" shouted Daniel, hoping to shake up Jurg. "You can't go to sleep. You must fight!"

But Jurg teetered on the brink of unconsciousness and was in danger of losing his life as the rescue crew chipped away at the ice slowly and carefully, so as not to further jeopardize his safety. Finally, five hours after he had first slipped into the crevasse, Jurg was lifted out.

"He was pale as death," recalls Christian. "His eyes were half-closed, his lips were deep purple. The freezing was taking control."

The group heaved a sigh of relief as the helicopter departed with Jurg for a hospital in Visp. There Jurg was treated for hypothermia and released three days later with no lasting effects.

"You have to be really lucky to survive falling into a crevasse," says rescuer Kurt Lauber. "Even if you fall a couple of feet, you can die."

Lauber praises Jurg's friends for remaining calm, which enabled them to think clearly and work together. According to Daniel teamwork is key in mountaineering.

"They constantly encouraged me," says Jurg. "They inspired me not to give up. They actually risked their lives for me. Without these people, I wouldn't be here today."

"We had no thought about being heroes. It was just the right thing—and afterward, I felt like Clint Eastwood," laughs Daniel.

Jurg isn't sure he'll go ski touring again in the near future, but Christian doesn't want people to get the wrong impression about the sport.

"All in all," he says, "a ski tour is surely less dangerous than driving a car."

Freeway the Fawn

On May 19, 1991, in San Diego, California, off-duty Sheriff's Deputy Lisa DiMeo was driving on Interstate 805, a major Southern California artery. Traffic was unusually heavy for a Sunday, and the cars were moving fast. Suddenly, a doe bounded out on the freeway and was struck by a car in the lane next to Lisa. The doe went down, but the motorist, traveling at high speed, continued on. The doe was lying between the middle lanes of traffic, still alive, and Lisa was worried that if the animal tried to stand up, she would be hit again. Lisa pulled to the shoulder of the road and called for help over her police radio. Her call came into the San Diego Sheriff's Communication Center, which dispatched a California Highway Patrol unit to the scene. Lisa also contacted the California Department of Fish and Game, but they would not be available for one hour.

Gene Chaffin was driving southbound on I-805 when he spotted something lying in the roadway. He thought it was a large cardboard box, but as he got closer, he saw that it was a deer. Gene put on his flashers and stopped his truck in front of the deer to protect her from other motorists. Lisa stayed on her radio and relayed events to the dispatcher while Gene and another motorist moved the doe to the shoulder of the road. The animal was in bad shape and was suffering. It appeared that she had a compound fracture to her right leg.

Within minutes, Highway Patrol Officer Greg Mullendore arrived. He and Gene Chaffin, who had grown up on a farm, agreed with dismay that the most humane thing to do was to put the deer out of its misery. Mullendore had never shot an animal before, and he was glad when it was over. Putting the doe down was an upsetting experience for everybody.

After the deer died, Gene noticed that her stomach was moving, as if something inside it were kicking. It dawned on him that the doe was pregnant. Suddenly his focus shifted from ending the doe's life to trying to save her fawn. Gene knew it was only a matter of minutes before it would die of oxygen deprivation. He ran back to his truck and grabbed a utility knife. He returned and cut open the doe's belly and delivered a fully formed fawn. Unfortunately, although it had a pulse, the fawn wasn't breathing.

Gene had never administered CPR to an animal, but he knew that time was running out. He performed "mouth-to-snout" resuscitation, and, within minutes, the fawn started breathing on its own.

"When he took his first breath, it was a great feeling," recalls Gene. "It was—yeah! I thought, wow, we saved him. It was a once in a lifetime uplifting feeling."

As Gene managed to get a little milk down the fawn's throat, someone proposed the name Freeway. Everybody chuckled, but the name stuck. The fawn would be called Freeway. Lisa drove Freeway and Gene to a nearby exotic animal hospital in her van. At the hospital, Freeway was cleaned up and its cord was tied. Its "foster father," Gene, was honored with cutting the umbilical cord. Gene and Lisa glowed with satisfaction that Freeway had survived.

The next day, Freeway was moved to the home of a zoo volunteer, who is raising him

in her backyard with the hope that he will eventually be released into the wild. When Gene and Lisa visited Freeway, he was very healthy and active.

"Gene did something remarkable out there," says Lisa. "Something beyond what most people would have done. To me he's a hero."

Gene thinks Freeway is incredibly lucky to be alive.

"This whole experience was probably the most rewarding thing I've ever done," he says. "Something like this happens to you only once in a lifetime. It was definitely a wonderful experience."

River Tubing Rescue

On July 26, 1989, near Chico, California, fifteen-year-old Reed Taylor and his friends planned to spend the hot summer day tubing on the Sacramento River. The kids brought an ice chest along and set it to float on its own inner tube. The tube kept floating away, so Reed, who lay on an inflatable raft, tied the ice chest to his ankle with a long nylon cord.

One of the girls in the group would warn the others of upcoming "snags," fallen trees in the water that presented a potential menace because of the powerful currents swirling around them. The kids approached a snag and kicked to avoid it, but Reed and eighteen-year-old Sue Miller, who were bringing up the rear, bumped into it. Sue fell off her tube, and the current sucked her under. She popped to the surface and yelled to Reed that her foot had become tangled in the limbs underwater. Reed swam over, freed Sue's foot, boosted her onto his raft and said he'd catch up.

But Reed didn't realize he'd swam into a death trap. The ice chest tied to his ankle had become stuck in the tree limbs and the rushing current was dragging him under. As he fought to keep his head above the surface, he tried in vain to untie the knot around his ankle or at least grab the snag to stay afloat. A passing cyclist heard Reed's cries and rode to get help.

Word of Reed's predicament reached Scotty's Landing, a dock and restaurant owned by John Scott, who on his own patrolled the river and rescued boaters and tubers. Scott jumped into his boat with two employees, including Eric Olson, and sped to the snag.

Meanwhile, Cindy Waters and a friend heard Reed's cries as they floated by on inner tubes. Cindy paddled to the bank, backtracked through thorny brush, and climbed on the snag. By the time she reached Reed, he was submerged, having sunk from exhaustion.

"I saw the life go right out of him," recalls Cindy. "I wanted to do whatever I could to do help him. But what I did wasn't good enough."

Cindy reeled in as much rope as she could with one free hand and her teeth, but she couldn't reach Reed.

Within moments, John Scott arrived. Olson leaned overboard, grabbed the rope and dragged Reed's body to the surface.

"The kid couldn't have weighed over one hundred-fifty pounds," recalls Olson, "and it took three guys well over two hundred pounds to get this kid out. The current was like a magnet."

Reed, who had been underwater for about twenty-five minutes, had no pulse and wasn't breathing. As the men motored back, they administered CPR, and Scott radioed for rescuers to meet at his dock.

Upon reaching their truck, Reed's friends heard that he was being given first aid at Scotty's, but they had no idea he was in serious condition. Meanwhile, people had gathered on the dock and some aided in CPR.

Paramedics Jeff Moore and Carey Allen arrived to treat Reed. They took one look at their lifeless victim and were immediately fraught with hopelessness.

Recalls Moore, "In the back of my mind I thought we don't have any hope, and maybe we're even doing this because there's a bunch of people watching us."

Reed's friends arrived and lapsed into a state of shock.

"I looked and there was nothing there," recalls Sue. "He was blue. It looked like that couldn't be Reed. It was hard for me to think . . . If anything happened to Reed—he saved my life."

When Reed was admitted to Enloe Hospital in a severely hypothermic state, he had regained a weak pulse and was breathing on his own, but he was otherwise clinically dead. The only thing Reed's parents, Judy and Rocky, and his twin brother, Jeremy, could do was to wait. For Jeremy, the situation was overwhelming.

"I tried thinking about Reed not being there and I couldn't," says Jeremy. "I couldn't even imagine that we wouldn't be together."

Around three in the morning, Reed opened his eyes. Remarkably, four days later, he was released from the hospital with no ill-effects of his near-drowning. Reed's physician, Dr. Terry Fraters, says Reed survived in part due to immersion in 55-degree water, which greatly reduces the amount of requirements needed to sustain life, especially the brain.

"But," says Dr. Fraters, "I feel that there was miraculous resuscitation. And if the Resurrection was a ten, this would score a nine."

"We'd be putting flowers on some stone someplace," says Rocky. "if just one guy had said, we've done all we can. It was an incredible chain of people that contributed."

Gator Gulch

On August 2, 1991, eight-year-old Brandon Herndon and his five-year-old sister, Ashley, were visiting their aunt, Vernell Waldrop, in rural Refugio, Texas. That afternoon, the children decided to swim in the Mission River, which flows behind Vernell's house, something they'd done countless times before.

Ashley played on the dock, while Brandon swam across the river in his life vest, anxious to show off his swimming ability to his aunt. Next door, Chris and Bob Stapleton's dog barked at Brandon from the back porch. Chris walked outside to hush her dog and noticed a log floating in the river toward Brandon. It took another second for Chris to realize that the log was actually an alligator.

"Bob, get your gun!" yelled Chris to her husband who was indoors. "There's a gator got a kid out here!"

Chris hollered to Vernell that there was an alligator in the water near Brandon. When Vernell looked up from the dock where she was playing with Ashley, she saw Brandon thrashing around in the water. The alligator had clamped his massive jaw on Brandon's arm. Brandon screamed for help, hit the alligator with his free hand and kicked him with his feet, but he couldn't free himself. Vernell dove into the water and swam toward Brandon.

"If the gator was going to take Brandon," recalls Vernell, "he was going to have to take me with him, because I wasn't going to give up."

Meanwhile, Bob ran down to the riverbank and aimed his rifle. He could not shoot the alligator yet without risking hitting Brandon.

"All I could do was aim, get ready and set," recalls Bob.

As Vernell swam across the river, the alligator released Brandon's arm. Brandon swam away, but the alligator attacked again, this time locking its jaw onto his leg. Brandon kicked and fought, but the alligator repeatedly dragged him underwater. Suddenly, the gator released Brandon for the second time. Brandon swam furiously toward Vernell.

Bob fired and the alligator sank out of sight.

Vernell reached Brandon, grabbed his life vest and pulled him to the bank. Chris kept a watchful eye on the water and saw the alligator resurface. He fired again and the gator disappeared for good.

"Aunt Vern, am I gonna die?" asked Brandon as his aunt took off his life vest.

Vernell explained that he was going to live and that she was going to take him to the hospital. Vernell called the hospital to alert them she would be arriving with a victim of an alligator bite.

The report of an alligator attack surprised the local game warden because as a rule, alligators are non–aggressive animals who do not attack unless they are cornered or are protecting their nest.

Brandon was treated at the hospital for multiple puncture wounds to his leg and arm and was released the same night.

"I'm proud of myself," says Brandon, "because I beat up an alligator."

Chris believes that Brandon's life vest played a major role in his survival by keeping him from being dragged under the water.

"If he hadn't had that life jacket on," says Chris, "he would have been in a world full of trouble."

Says Brandon, "I will never, ever, in my whole life get back in that river till I'm thirty-eight."

Idaho Trooper Down

On the night of June 15, 1991, Idaho State Trooper Corporal Steven Hobbs was on routine patrol near Sweetzer Pass, a remote stretch of highway near the Utah border, when he spotted a speeding car. Hobbs flashed his lights and the car pulled onto the shoulder and stopped. Hobbs approached the car on foot and barely got the words "Good evening" out of his mouth, when the driver fired at him four times with a nine-millimeter semiautomatic pistol. The one bullet not deflected by Hobb's bullet-proof vest entered his left shoulder, exited his body, and re–entered his right arm.

Around 9:15 P.M., police dispatcher Dorinda Silver heard loud static over her radio. She suspected it might be an officer trying to radio from Sweetzer Pass, an area known to police for poor or no radio transmission. Then a voice crackled over her radio.

"I need help, 475, I need help." Hobbs's I.D. number was 475. Then Dorinda was sure she heard the word *shot*. She couldn't contact Hobbs, so she dispatched an ambulance to the general location. She also notified the nearest officer, who was seventy miles away.

Meanwhile, Hobbs tried to pursue the driver. But after one mile, slipping in and out of consciousness, he lost control of his car and came to a stop on the highway's sage-brush-filled median. His engine was still running, and within minutes the dry brush beneath his car caught fire.

Floyd Veibell and his family were on their way home in two cars from a family reunion when they saw Hobbs's patrol car engulfed in smoke. Floyd and his family stopped to see if they could help. With his family fearful that the burning car would explode, Floyd ran over to see if someone was inside it. He opened the door and found Hobbs slumped over the wheel.

The Veibells laid Hobbs on the ground and found the bullet holes in his arm and shoulder. Floyd's daughter-in-law, Sherri, noticed Hobbs's breathing was shallow. A moment later it stopped. Although Sherri wasn't trained in rescue breathing, she gave it a try. Three times she performed mouth-to-mouth resuscitation, and each time Hobbs began to breathe, then stopped. Sherri and Floyd worried that they were going to sit there and watch the man die.

"You feel so absolutely responsible," recalls Floyd. "You almost want to say, you son of a gun, you better not die on me." Floyd felt vast relief when he spotted the two men running toward them carrying a black medical bag.

Off-duty EMT John Cook and his brother, EMT trainee Bill Dennison, happened to be on their way home from vacation when they spotted the fiery wreckage. John made an assessment of Hobbs and knew he had to get him to a hospital quickly. Hobbs was in deep shock and combative, which is the body's response to oxygen deprivation of the brain. Since they couldn't radio for help, John instructed one of the Veibells to flag down an approaching mobile home. The driver agreed to transport Hobbs, and rescuers Floyd, John, and Bill, to Snowville, ten miles away. The men worked to keep Hobbs alive until he could be airlifted to McKay-Dee Hospital.

Meanwhile, Hobbs's wife, Janice, a Minidoka County dispatcher who was on duty that night, heard about the accident over the radio, and was rushed to the hospital by a police officer.

Hobbs arrived at McKay-Dee, two hours after he'd been shot. Following successful surgery to repair the severed artery and nerves in his arm, Hobbs underwent physical therapy, which he still continues.

Today, Hobbs has limited use of his hand, and as a result of the oxygen deprivation to his brain, has partial short-term memory loss, and a permanent 50 percent loss of vision. Unable to do many things he was previously able to do, including working on patrol, Hobbs has had to face a huge adjustment, but he's thankful to be alive. He and Janice are grateful to all the people who stopped that night to help.

"Steve means everything to me," says Janice. "I will never be able to repay the people who took the time to rescue him. I never knew there were so many people who cared."

Floyd Veibell's best reward was seeing Steve play with his seven children at a party thrown for his rescuers. Floyd was reminded of when he was a kid and happened to see the accident in which his own father died.

"So I was raised without a father," says Floyd, "and these kids had theirs. To me, that's all that counts."

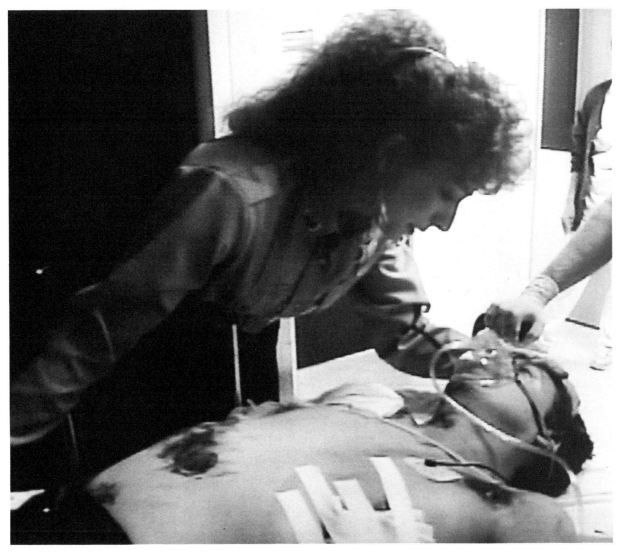

New Zealand Tanker Fire

On the evening of August 9, 1990, Gaylene Young of Manukau, New Zealand, stopped her car in the road to drop off her twelve-year-old daughter, Shirley, at the local mall. David Petera was standing in the parking lot when he noticed Shirley run back to the car and lean inside to talk to her mother. That moment, not far up the road, a truck hauling more than 8,700 gallons of gasoline almost collided with a car that pulled out in front of it. The trucker swerved and missed the car, but he didn't see Gaylene stopped in his path. He turned the wheel sharply to avoid Gaylene—but not in time. The tanker slammed into the car, jackknifed, and crushed Gaylene and Shirley underneath it. Then both vehicles burst into a fiery ball.

Rescue units from the Manukau Fire Department sped to the scene as Petera and the trucker, who managed to escape, pulled a screaming Gaylene from the fiery wreckage. When Station Officer Graham Haycock and Senior Firefighter Royd Kennedy arrived, flames were shooting sixty feet into the air, and the heat was so intense that they couldn't get within thirty feet of the tanker. A river of fiery fuel poured from the vehicle

and headed straight for the mall. Haycock and Kennedy saw Gaylene lying on the ground, badly burned, and they hosed her off to cool the burns.

"My baby! My baby's in the car!" she cried.

The firefighters knew her baby had to be dead. Haycock and Kennedy went to the back of the tanker where they were engulfed in black smoke swirling at ground level. Suddenly, Kennedy could have sworn he saw an arm waving in the smoke. He ran straight into the fire where he saw, crushed under the wheel of the tanker, a little girl screaming for help.

"She was really scared," recalls Kennedy. "Mind you, so was I. But I let her know I wasn't going to leave her—no matter what."

At the risk of losing his own life, Kennedy wrapped his arms around Shirley and stayed by her side while firefighters battled the blaze. Fire flashed past Kennedy and Shirley, and freezing cold water beat down on them from the firefighters' hoses. "When you get out of here," Kennedy told her, "We're going horse riding."

"She was a battler," adds Kennedy. "And it's a great thing to draw your own reserves of strength from when you've got someone like that next to you—who's only twelve years old."

In a race against time, rescuers worked to boost the tanker's wheel assembly just high enough to lift Shirley out, using air bags and hydraulic pump. But Shirley started to fade.

"If I don't make it," she said to Kennedy, "tell my mother I love her."

Kennedy administered oxygen to a now unconscious Shirley until rescuers succeeded in raising the tanker. Forty minutes after being trapped under the vehicle, Kennedy carried her out of the wreckage, placed her on a stretcher, and gave her a kiss on the cheek. Then he collapsed himself.

"After the rescue was completed," recalls Haycock, "the fire just went out. Just like that—*foof*! And I felt as if the devil had backed off now that Shirley had been removed from the fire."

Shirley and Gaylene spent months recovering in the hospital. Shirley suffered second and third degree burns over 20 percent of her body. Both of her legs were badly broken, one so severely that it had to be amputated. Shirley made a marvelous recovery, and five months after the accident, Royd Kennedy kept his promise and took her horseback riding. Their relationship has grown into a very special one, and Kennedy thinks of her as a daughter. Gaylene feels that Kennedy's actions went way beyond heroism.

"If Shirley had had to go," says Gaylene, "I don't think he would have left her side. I think he would have given his life."

"Royd did what I hope any firefighter would have done." says Divisional Officer Ray Warby. "But I wouldn't ask any firefighter to do it."

"So many people can learn from that child and what she went through," says Kennedy, "from her strength of character. She's a million dollars, that kid. Boy. We haven't heard the last of her."

Runaway Carriage

On January 27, 1991, Wayne Shupe and his friend, Colleen Frey, took Wayne's horse and carriage for a ride on the back roads of Bushkill Township, Pennsylvania. Wayne had been training his horse, Spider, for one month to pull the antique carriage he had recently bought. Wayne turned off the highway onto a gravel road where Spider trotted toward Wayne's farm, four miles away.

Thick woods on Spider's right suddenly gave way to a field with a large bale of hay sitting by the road. The hay spooked Spider and he reared up and started to stamp his feet. Colleen jumped from the carriage to safety just before Spider took off in a panic and broke through an electric fence. This spooked Spider even more. He ran through the fence three different times, starting and stopping suddenly, causing Wayne to be thrown over the front of the carriage and get his right leg trapped in the spokes of the front wheel.

Colleen watched in horror as Spider galloped away with Wayne, who was straining to hang on and keep himself upright to prevent being crushed under the wheels of the carriage. Spider ran into the brush and Wayne was knocked unconscious. His horse dragged him a quarter-mile down the road, then into the woods, leaving a trail of broken trees in his wake. Colleen ran after Spider, screaming. She heard the initial crash of the carriage entering the woods, then she heard silence.

When Colleen reached Wayne, his leg was tangled in the wheel, and his clothes, which had been torn off his body, were wrapped around his throat. Colleen ripped the clothes from his neck. She couldn't untangle his foot, nor could she unharness the panicked horse by herself. Colleen had to get help fast because Wayne was having trouble breathing. It was also just a matter of time before Spider bolted again and dragged him to his death through the woods. Colleen ran for the nearest house.

Debra and David Seese and their children were out for a Sunday drive on a road they rarely traveled when Colleen frantically waved them down. David, not knowing what to expect, left his family in the car and followed Colleen into the woods.

"When I first saw Wayne," recalls David, "I thought he was dead. The scene was a mess."

David and Colleen jockeyed the carriage back and forth until David was able to untangle Wayne's leg and pull him out from under the carriage.

Meanwhile, Debra wondered what had happened to her husband and drove down the road and shouted to him. David yelled back to go for help. Debra drove to the nearest house to call for help. Members of the Bushkill Township Volunteer Fire Company, the local police, and medic units were dispatched to the scene.

Because of the threat of internal injuries, Wayne had to be transported to the hospital quickly. He was bleeding badly from his head, had a depression on his right chest, which signaled a punctured lung, and he was combative, a sign of severe head trauma. Wayne was loaded into the ambulance, and, less than ten minutes later, arrived at the hospital.

Spider escaped the incident with only minor cuts, but Wayne suffered a collapsed lung, three broken ribs, a cracked sternum, dislocated shoulder, and multiple leg fractures.

In spite of the accident, Spider is still Wayne's favorite horse.

"The happiest moment," says Colleen, "was when Wayne got out of his truck after he got home from the hospital. He went back to the barn and said, 'Hey, Spider, how you doing?' "

"My horse is a pal," says Wayne, "a good friend and buddy. Accidents happen. I will definitely get back in the carriage with Spider pulling it. I have no fear of that. If you fall off, you get back on again and ride."

Wayne knows he's lucky to be alive and credits rescuers for his survival.

"Everybody that came to my rescue was a fine, super crew and gave me the best medical attention they could. I have to say, they saved my life, and thank you."

Stop, Drop and Roll

On President's Day, February 12, 1990, eleven-year-old Leighton Oliver, his nine-year-old brother, Tristan, and a group of friends were playing "war" with squirt guns outside the Olivers' home in Aurora, Illinois. Marcia Oliver told her sons she had to run an errand and would be back in fifteen minutes.

Leighton soon tired of the game and thought it would be fun to build some small fires. Tristan and the other boys, including thirteen-year-old Jimmy Marrello, went along with Leighton's plan. The boys tossed leaves, twigs, and pine cones into a small pot while Leighton went into the garage. He poured a little gasoline from a tin can with a spout into a glass jar and carried it out to the yard. Leighton poured the gasoline into the pot and lit it with a match. Small flames erupted and the boys doused them with their squirt guns.

"Let's make a bigger fire," suggested Leighton.

He went back to the garage and returned with the gas can. Tristan and the other boys backed away as Leighton poured more gasoline into the pot. Instantly, the fumes from the gas can ignited, causing a small explosion and setting Leighton's clothes on fire.

Jimmy saw Leighton covered with flames, waving his arms in the air. Instinctively, he remembered the drill he'd learned in school—if your clothes catch fire, "Stop, drop and roll," then smother the flames with a blanket. Jimmy yelled at Leighton to drop on the ground and roll, then he ran to him and smothered the flames with a jacket.

Jimmy then ran inside and called 9-1-1, while Tristan took Leighton to the water faucet and soaked him in cool water. Leighton screamed in pain, fearful he was going to die.

"I just stayed with Leighton, crying my eyes out, and worrying about him," recalls Tristan. "I thought that he would die for sure."

Rescuers from the Aurora Fire Department, including paramedic Joe Bartholomew, arrived on the scene within minutes and began to cut away Leighton's polyester sweatshirt, which had melted and formed a plasticlike coating on his skin. As Leighton was loaded into the ambulance, he repeatedly asked paramedics if he were going to die.

Assistant paramedic trainee Tony Nelson felt badly that he couldn't do more to comfort Leighton.

"I couldn't hold him, couldn't hold his hands, couldn't give him anything to grab because his hands hurt so badly. All I could do was soak him down and tell him, 'If you feel like screaming, go ahead and scream.' He did."

Marcia and her husband, Tony, rushed to Mercy Center Hospital where Leighton was treated for burns over 60 percent of his body. Nurse Linda Hemmingsen recalls the mood in the trauma unit.

"We felt this tremendous sadness that he was burned so severely that he probably wasn't going to make it," she says.

Leighton was airlifted to Loyola University Medical Center's burn unit. There his family and friends began an intense vigil to see if his body could fight off infection, the greatest threat to a burn patient's survival. The hardest part for Marcia was not being able to comfort Leighton by touching or holding him.

Incredibly, Leighton slowly improved. He underwent skin grafts and endured a painful cycle of bandage changes and physical therapy.

Today, Leighton has recovered and has returned to all his usual activities, including his favorite, soccer.

"I don't know the name of his friend," says paramedic Nelson, referring to Jimmy Marrello, "but that's a friend for life. He saved Leighton."

A lot of people think Jimmy's a hero, but he says he just did the first thing that came to his mind.

"Absolutely he's a hero," says Leighton. "I probably wouldn't be here if it wasn't for Jimmy."

Marcia says if there's one thing she's learned, it's that gasoline should be locked out of the reach of children. Paramedic Joe Bartholomew agrees.

"Children have a curiosity about fire," he says. "The problem with playing with gasoline is that they don't understand what's going to happen. The vapors spread out and will ignite. There's no way they can put that can down before it will explode."

Best Buddy Rescue

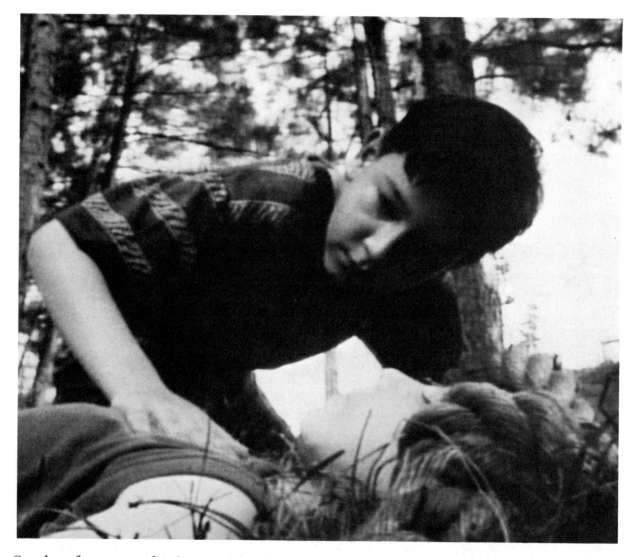

On the afternoon of February 24, 1991, Ryan Laurence and his friend Stuart Cary, of Ruston, Louisiana, planned to build a fort. Stuart's father, Bill, gave Stuart his hatchet so the boys could cut small trees and branches. Stuart's mother, Vicki, argued with Bill about the wisdom of allowing eleven-year-old boys to play with a hatchet, but Bill was convinced Stuart was old enough and responsible enough to handle the tool.

For a couple of hours Stuart and Ryan worked on their fort in a pine thicket at the elementary school, which was a few blocks from Stuart's house. On their way home, Ryan saw a rope hanging from a tree that the children used for swinging. The rope, which had a loop at its end, hung a few feet above the ground at the edge of a hill. Ryan, a practical joker, ran to the rope and slipped the loop around his neck.

"Hey, I'm going to hang myself," he yelled to Stuart.

"That's not funny," Stuart replied.

Stuart heard some kids in the distance and turned to see who they were. Moments later, when he turned back to Ryan, he saw him hanging limply from the rope. It appeared that his knees were resting on the ground.

"Hey, that's not funny," Stuart yelled again, annoyed.

Stuart went over to Ryan and shook him, but Ryan didn't respond. Slowly it dawned on Stuart that Ryan wasn't joking. In fact, he was unconscious. Ryan must have lost his footing on the slippery pine needles. The incline of the hill had kept his feet from touching the ground.

Stuart tried to remove the rope from Ryan's neck, but Ryan's weight made the noose too tight. Stuart was scared his friend was dead, but he told himself to stay calm. He ran for the hatchet, which he'd dropped nearby. Stuart wasn't sure if he was doing the right thing, but he decided to cut through the rope with the hatchet.

Once he sawed through the thick rope, Ryan fell to the ground. Stuart ran to his house for help.

"Come quick! Ryan's hurt!" he screamed as he ran up the driveway.

The Carys drove a few blocks to the thicket and found Ryan convulsing. He was blue and had a terrible rope burn around his neck. Vicki ran to a neighbor's house and dialed 9-1-1.

Captain Stephen Beard, who was patrolling a few blocks away, arrived moments before EMT Captain Kenneth Ambrose and his rescue team. Ambrose immediately recognized Ryan—his own son went to church with him.

The medics had to move Ryan down the hill while keeping his airway open—no small feat, since the bank was steep and slippery. Ryan's mother, Vané Geter, arrived. As she watched the medics load Ryan into the ambulance, she thought she felt him slipping away.

Upon his arrival at Lincoln General Hospital, Ryan was given medication to control his seizures. Then he was airlifted to Schumpert Medical Center in Shreveport, where he was treated. The doctor told Vané that they would have to wait to see if Ryan came out of his coma without brain damage. The big unanswered question was how long Ryan had been deprived of oxygen: that would be key in assessing the extent of brain damage he might suffer.

Vané sat by her son's side. After sixteen long hours, Ryan opened his eyes and turned to his mother.

"Ryan, it's mama. Do you know who I am?" Vané asked.

"Vané," he responded in a tone that implied it was obvious.

Two days later, Ryan was released from the hospital. He completely recovered and suffered absolutely no brain damage.

Since the accident, Stuart and Ryan have become better friends.

"Stuart and I are closer now that this happened," says Ryan. "I hope I'll be his friend for the rest of his life, and I'll always be there if he ever needs me."

Vicki and Bill are proud of their son for keeping a cool head and acting quickly. If he hadn't, Ryan wouldn't be here today.

Vané is also proud and appreciative of Stuart.

"I thank him and love him from the bottom of my heart. Our families will have a bond that will never be broken."

9-1-1 Market Robbery

A little after midnight, on April 18, 1992, nightshift employees of the Bel Air supermarket in Roseville, California, were closing the store and balancing their cash registers before going home. The automatic electronic doors had been turned off but had not yet been locked.

Employees Chad Graham and Pat Maldonado said good night to their co-workers as they were leaving the market. Seconds later, four masked armed men forced their way through the electronic doors and ordered employees to lay facedown on the floor.

Chad caught a glimpse of the gunmen as he was exiting. He ran to a pay phone at the front of the store and dialed 9-1-1.

"There's a holdup in Bel Air on Sunrise and Kirby! Quick!" he told the dispatcher.

It was dispatcher Pamela Hardwick's first robbery call in four years of service. She was nervous as she dispatched police to the scene, because she knew anything could happen.

Inside, terrified employees were neither moving nor obeying the gunmen's orders. Head clerk Ken Pickard got down on the floor and yelled for his employees to follow. While one of the gunmen guarded the employees, the other three ran to the customer service office. A gunman pounded on the office window with his gun, demanding to be let in. Employee Tracy Eagelston, who had been hiding under a desk, came out and let the men in.

Stephanie Soulier lay on the floor and heard Tracy scream as the men rushed in and ordered her to open the two safes.

"I was so scared when I heard Tracy scream," says Stephanie. "She's like my little sister. It was as if somebody in my family was going to get hurt."

Meanwhile, Chad stayed on the phone, updating Hardwick as best he could about the gunmen's movements. He was afraid that the driver in the getaway car in the parking lot would see him.

"I'm in front of the store," Chad told Hardwick, "and I don't want them to see me. Oh God."

Within forty seconds of being dispatched, police units arrived on the scene with their lights off. Chad got off the phone with Hardwick and stood by as Watch Commander Rocky Rockholm instructed officers to surround the building.

"I was scared to death," says Rockholm, who had been the supervisor of the night shift for only four nights. "I thought of all the things that could go wrong. I said a little prayer to myself."

Inside the customer service office, gunmen took the cash from the top safe and demanded that Tracy unlock the bottom one. She told the gunmen that only Ken Pickard had the combination, so they ordered Ken into the office to open the safe. Ken told the men there were only coins inside, but the robbers were determined. Ken stalled with the combination, but he finally opened the safe, which held stacks of rolled coins and a little cash.

Angered by their small take, the gunmen ordered Ken to lay on the floor. One of the men cocked his gun, put it to Ken's head, and ordered him to count to ten. Ken slowly counted out loud, waiting to be shot. Instead, the men fled.

The suspects ran out of the market directly into the path of two police officers, who shouted, "Police! Freeze!" Officers immediately apprehended two of the gunmen, but two fled on foot. A pursuing officer caught one of the suspects in a nearby field. Chad saw the fourth suspect running away and yelled to police. A police dog was released and tackled the man to the ground.

Rockholm credits good police work to the fact that nobody was hurt and the suspects were captured. Chad Graham feels lucky that his exit was perfectly timed. Had he left a moment sooner, he wouldn't have seen the gunmen. Had it been a moment later, he would have been trapped with the other employees in the market.

Store employees were shaken by the robbery, but they returned to work the next night with tighter security.

Ken Pickard is happy he survived to see the birth of his daughter two months later.

"When you look in her eyes, it just tears your heart out that you might have missed something like that," says Ken. "I'm happy that I'm alive to see my little girl."

EXTRAORDINARY STORIES

Storm Drain Girl

On June 26, 1989, tropical storm Allison blew into Texas and dropped ten inches of rain in one day. By seven o'clock that evening, after the storm had stopped and water was being redirected into Houston's storm drains and bayous, Karen Reese allowed her daughter, Tricia, and niece, Krystal, to play outside in the flooded street. The water didn't look dangerous, but within moments, seven-year-old Tricia was dragged down by the rushing current and sucked into a storm drain. Krystal ran inside and told her Aunt Karen what had happened. Karen called 9-1-1.

Firefighters from the Houston Fire Department arrived on the scene with Fire Captain Bob Crenek in charge. One of Crenek's men jumped into the water at the opening of the eighteen-inch-wide drain to see if he could feel anything inside. He emerged empty-handed, saying that if he hadn't been so big, he, too, would have been sucked in by the swift current.

The rescuers also checked for Tricia directly across the street, where water was rushing at forty miles per hour along a drain gutter. There was no sign of Tricia.

"I felt so helpless," recalls Karen Reese. "The hardest part was standing there watching them working at that drain. It just seemed so futile because there wasn't anything anybody could do."

A police dive team was called in to search the bayou at the outlet of the storm drain. For three hours, a boat traveled the length of the nearby bayou, as men walked up and down the banks. By nightfall there was still no sign of Tricia. Police called off the search until the next day, doubtful she would be found alive.

Early the next morning, before the police dive team resumed their search, Tricia's fifteen-year-old cousin, DeVincent Phillips, set out to look for Tricia one more time. A few blocks from the Reeses' home, DeVincent came across two men who were looking at blueprints. Tim Gabrysch and his partner were checking sewage pipes in the area for storm damage, but DeVincent assumed from the blueprints they might be looking for Tricia. Gabrysch and his partner had, however, heard about the little girl's disappearance over the radio and agreed to help. Although Gabrysch couldn't imagine finding anyone after all the water that had raged through Houston's streets and drains, he agreed to follow DeVincent to the site of the accident.

"The first thing that went through my mind," recalls Gabrysch, "was that I'd rather not find anything than find a dead little child."

By now, the flood waters had receded enough for Gabrysch to climb down the nearby manhole. He aimed his flash light down the length of the large pipe and was stunned by what he saw. A little girl was huddled in the pipe. Tricia had clung to the wall of the three-foot-wide drain for thirteen hours, and she had survived. Gabrysch called out to her, but she was hesitant to come out.

"I want my mama," cried Tricia.

Gabrysch replied that her mama had sent him and he was going to take Tricia to her.

Karen heard the news and ran down the street to meet Gabrysch and Tricia. She swept her daughter into her arms and both of them cried in joyful reunion.

Gabrysch was elated and shocked that Tricia had escaped her ordeal unscathed.

"You could have another million people fall in like Tricia and I don't think any would live like she did. I don't know how she held herself against that current."

Since the accident, the city has installed grates at the opening of the drains. Tricia still plays outside, but now she stays in her backyard.

"I play in the back because I can still remember what happened," she says. "I missed my mommy. And my hands, they were wrinkled and all white. They weren't like my human hands."

Karen says that everyone in the family always protected Tricia more than the other kids because she seemed so frail. Not anymore.

"I tell Tricia, you went down that drain yourself and you came out of it okay yourself. So don't tell me you can't do anything by yourself because you've done this," says Karen. "Thank God he gave her that miracle."

Truck vs. Train

Linda and Ronald Schams worked hard over the 1991 Memorial Day weekend in their bait and boating shop in Onalaska, Wisconsin. By Monday night, the only thing on their minds was relaxing at home.

Around midnight, the Schams were awakened when the burglar alarm in their shop rang. The alarm system had been installed only recently, due to a series of burglaries at the store, which was located just across the railroad tracks from their house.

While Linda dialed the Onalaska Police Department to report the break-in, Ron jumped in his pickup truck, determined to catch the thief in the store. As Ron drove toward the railroad crossing, he saw a train coming down the tracks. Ron was in a hurry. He didn't want his burglar to get away, but if he stopped for the train, that's exactly what would happen. So, Ron went for it—figuring he could make it across the tracks in time to beat the train.

But Ron misjudged the train's speed and distance. As he sped across the tracks, the train slammed into his truck. On impact, the truck flipped over and Ron was ejected from the rear window. He landed in a ditch and the truck landed on top of him.

The conductor, who had witnessed the accident stopped his train and called 9-1-1. Units from the Onalaska Fire Department and Tri-State Ambulance were immediately dispatched to the scene. Just as Linda went out on their deck to see if Ron had reached the shop, the phone rang. It was a 9-1-1 dispatcher calling back to ask her about the train wreck. Linda had heard the train whistle blowing, and the burglar alarm was still ringing in the shop. She immediately put two and two together and knew Ron had been hit by the train.

"My whole life seemed to stop right there," recalls Linda. "I thought, how do you get out of a truck hit by a train without being in pieces?"

Within minutes, police officers and rescuers arrived.

"Get me out of here!"

Paramedic Bruce Solberg heard the anguished scream that came from beneath the truck. Solberg kneeled under the flatbed and had introduced himself before he realized the victim was someone he'd known for twenty years. Paramedic Dave Blocker, who also knew Ron, crawled under the truck to assist. Blocker felt Ron's ribs and realized the situation was critical. His ribs were all over the place, indicating that Ron's lungs could collapse.

Rescuers were posed with a dilemma. Ron needed to be extricated immediately, but first the truck had to be removed safely. A tow truck arrived on the scene but was unable to reach down far enough to lift the truck out of the ditch. Another big concern was that the truck might shift while being lifted and fall back on Ron and the paramedics, killing all three.

Solberg and Blocker were advised to come out from under the truck, but they refused, saying they were not going to abandon Ron. By now, Ron's right lung had collapsed and his condition was rapidly deteriorating. Blocker and Solberg, who rated Ron's

chances of survival at 5 percent, knew they were out of time. Blocker decided he would have to try lifting the truck with his own hands. With the aid of several firefighters, Blocker positioned himself under the truck and raised the flatbed just high enough for two paramedics to pull Ron out.

Ron was transported to the hospital, where various tests revealed eighteen broken ribs, two punctured lungs, a cracked vertebra, and a shattered shoulder blade. Remarkably, he suffered no major internal injuries.

Ron and Linda were so grateful to everybody involved in saving his life, that after Ron's recovery, they threw a thank-you party.

"With the injuries that Ron incurred," says Solberg, "we should have been going to his funeral instead of a party."

Having a second chance at life has changed Ron's outlook.

"Anger blinds people," he says. "I had one thing on my mind and that was getting the bad guy. I thought I could beat the train and I didn't rationalize what could happen. I didn't think it through."

"There must have been a guardian angel looking out for Ron," says Linda, "because you don't normally tackle a train and come out on the winning side."

9-1-1 Sucker Save

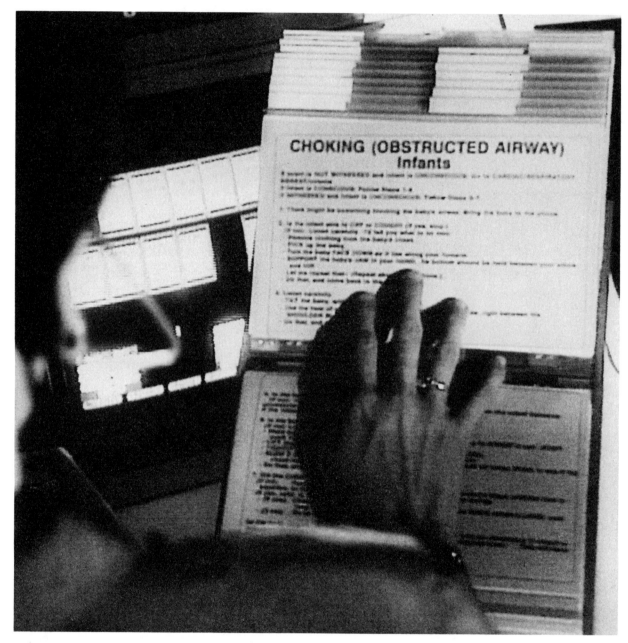

Lisa Smith, a twenty-seven-year-old mother of four children, works as a babysitter in her home in Indianapolis, Indiana. On the afternoon of June 5, 1992, Lisa was looking after sisters Melanie and Mindy Cox. While her own children played in the backyard, Lisa read to two-year-old Mindy and her four-year-old sister. When she finished the story, Lisa gave each girl a lollipop.

Lisa turned her back on the girls to throw the candy wrappers in the trash. At that moment she heard a choking sound and turned around to see Mindy gasping for air.

Lisa ran to Mindy and looked in her mouth. She could see the lollipop lodged in the back of Mindy's throat, but she couldn't reach it because the stick had apparently broken.

Lisa dialed 9-1-1 and dispatcher Carol Wash answered her call. Lisa hysterically reported that Mindy was choking. Wash told Lisa to hold on and immediately dispatched Indianapolis Fire Department rescue units. When Wash returned to the line, Lisa was still hysterical and told her that Mindy was barely breathing and turning blue.

"Stay calm," said Wash. "You have got to stay calm."

Wash turned to her emergency medical dispatch cards, which outline step-by-step instructions on performing emergency procedures. In her three years as a dispatcher, this was the first time she had needed to use the cards.

"My insides were just going crazy," recalls Wash. "You've got four to six minutes to do what you've got to do. After that, it's real touchy."

Lisa followed Wash's instructions, giving back blows to Mindy. On the fourth blow, the sucker dislodged from Mindy's throat.

The Indianapolis Fire Department arrived on the scene within three minutes. The EMTs knew Mindy was all right when they heard her crying loudly. EMT Mark Oster checked Mindy's throat and breathing. She was fine.

Lisa says she had no idea how dangerous lollipops could be to small children. Since the incident, she does not keep them in the house, nor does she allow her children to have them.

Lisa tearfully recalls that day. "I can't describe the terror I was feeling. I love Mindy like she were my own, and I felt guilty that I was responsible for her choking. I thought I was going to lose her for sure that day."

Mindy's parents are thankful for Lisa's actions, which saved their daughter.

"If not for 9-1-1 and Carol Wash," says Lisa, "Mindy wouldn't be here today."

Laundry Chute Lad

Dr. Mary Neuffer of Columbia, South Carolina, had always been extremely conscious of the accidents that usually befall children, but on May 3, 1992, her son, John, fell victim to one that was unimaginable.

When Dr. Neuffer was called to the hospital to deliver a baby, her sister-in-law, Rene Neuffer, came over to babysit seven-year-old John and his sisters, four-year-old Katie and two-year-old Caroline. Rene and the kids were playing a game of hide-and-seek, and the children scrambled as their aunt counted to one hundred with her eyes closed. John, who had earlier grumbled about not finding any good hiding places around the house, ran into his parent's bedroom.

John scanned the room. There it was, he thought, the perfect hiding place. No one would find him in his parents' laundry chute. John opened the door and climbed inside the six-inch-by-twelve-inch chute. He intended to balance on a small ledge at its mouth, but he immediately lost his footing and slipped, feet-first. John's body was now wedged inside the narrow shaft, halfway between the first and second floors.

Rene was still counting when she heard John's cry for help. She ran into the bedroom and looked down the chute, where she saw the top of John's head and one small hand. She reached in and tried to grab it, but he slipped farther down. Rene realized she needed help, so she phoned her mother, seventy-two-year-old Irene Neuffer.

Irene drove over, took one look at the situation, and decided the only way to get John out was to cut the frame of the chute. They got the tools, but the vibrations caused by the sawing made John scream in fear and pain.

"Oh, Meema, just call 9-1-1" John called out.

"Bless your heart. Why didn't I think of that?" responded Irene.

By the time rescue units from the Columbia Fire Department arrived on the scene, John had been stuck in the laundry chute for one hour. Firefighters, not trained in this type of rescue, discussed the best approach to extricate John, while paramedic Britt Ogden tried to loosen him with a slick substance called "defib jelly." It soon became evident that John wasn't going to budge, so the firefighters concurred they would have to rip apart the ceiling in the kitchen, where the chute emptied out. While the men worked slowly and carefully, John grew more frightened, sure that he was going to die.

"I was saying to myself, I'm gonna be stuck here forever," recalls John. "And I didn't want to die younger than Elvis."

Two hours after John got stuck, firemen had finally ripped away enough of the ceiling and the shaft to pull him out, into the kitchen. The process had been a slow one to insure John's safety. John insisted that all he wanted was a Tylenol and rest, but paramedics didn't want to take any chances in the event of skull or neck injuries. They put him in a C-collar, strapped him to a backboard, and transported him to the hospital, where his mother was delivering a baby.

Dr. Mary Neuffer met John in the emergency room and was relieved. He was covered with jelly and soot, but otherwise he was in healthy condition.

When John returned to class, he recounted his narrow escape to a school assembly. He wanted his schoolmates to know that it's important not to hide in small places.

"Don't go in small places with walls around it where you'll get stuck," said John. "And never go near a laundry chute unless you are as skinny as a stick."

Irene is pleased that her grandson came out of the adventure unscathed.

"John Henry, he's the pick of the litter," says Irene.

Blind Hero

Around two-thirty in the morning on December 31, 1987, thirty-four-year-old Felix Bannon was returning to his Northern California home after a short trip to the corner store. Felix, who's been blind since birth, has always been independent. He shares his apartment with his guide dog, Valdez, and a lot of musical recording equipment. Playing and recording music is Felix's greatest joy in life and his apartment is a virtual mini-recording studio.

Once inside, Felix turned on the space heater to warm his bedroom on this very cold night. Valdez lay down on the floor, and Felix got into his pajamas and called a friend. As Felix sat on his bed, and talked on the phone, he heard the heater shut off. Thinking this was strange, since the heater hadn't been on long enough to warm the room, Felix decided to investigate. He crouched down, reached under a table, and unplugged the heater. Then he heard a rush of fire inches away from his face.

Felix grabbed the phone and dialed 9-1-1. He reported the fire and his address to the dispatcher, then he dropped the receiver. As the fire built to a roar, Felix ran outside with Valdez, with only one thought on his mind—his neighbors' safety. Most of the fourteen other tenants in the six-unit building were asleep, and Felix had to warn them to evacuate because the wood-shingle building they shared would go up in flames quickly. Wearing only his pajamas and socks in the subfreezing night air, Felix raced from door to door to alert his neighbors, some of whom he barely knew.

Felix was especially worried about two neighbors downstairs. One had Cerebral Palsy and would have to be lifted out of bed and into his wheelchair by his attendant. The other, who lived directly below Felix, was a very heavy sleeper. Felix was concerned the fire would burn through his own floor into the man's apartment before he would be able to wake him. As Felix heard windows exploding upstairs from the heat, he ran around the side of the building and yelled into his neighbor's bedroom window. The man finally woke up and escaped unharmed.

Within three minutes of Felix's emergency call, the Novato Fire Department arrived and started to put out the blaze. Firefighters say that if they had arrived three minutes later, they might have lost half the building.

Novato Fire Marshal Tom Elliott found Felix in a nearby apartment building, shaken and in need of oxygen to clear his lungs. Despite Felix's condition, Elliott remembers being struck that he repeatedly asked questions about his neighbors and never once asked about his own belongings.

The investigation into the cause of the fire revealed that it started as a result of somebody else having spliced the heater's electrical cord. Felix lost all his possessions, but because of his quick response, nobody got hurt and only his apartment was seriously damaged.

"The one thing I always remember from the fire," says Felix, "is that nobody got hurt, and that meant a lot more to me than the things I owned."

Since the fire, Felix's apartment has been rebuilt and he still lives there today. He's put his life back together, thanks to some help from a local music store that held a fund-raiser to replace his instruments and recording equipment.

"It makes you appreciate every day to have what you love and the people you love around," says Felix. "When people ask me how I did what I did, I tell them being blind wasn't a disadvantage. It wasn't an advantage, but it was just a matter of dealing with what had to be done. The costs could have been immense for lots of other people besides myself."

In recognition of selfless courage, the California State Firemen's Association awarded Felix with a medal of valor, the first one bestowed upon a blind person.

It's a Boy

On April 25, 1992, Beth and Mike Mannino of Lawrenceville, Georgia, were eagerly awaiting the birth of their second child, due in one month. Around six-thirty in the evening, Beth started having painful contractions, and asked Mike to run to the store for aspirin and juice. As Beth lay on the bed, her contractions became more frequent and unbearably painful. Then her water broke. Frightened that she might go into labor alone, she called 9-1-1.

"I need an ambulance," Beth told Gwinnett County Dispatcher Bob Cooper. "I'm thirty-six weeks pregnant and my water just broke, and I think I'm having it—now."

Beth explained that she was home alone and that her first baby had been born thirty minutes after her water broke. Cooper grew tense. This was his first obstetrics call, and he immediately turned to his medical information cards, which give step-by-step instructions on how to handle any given emergency.

"The answers she was giving me to my questions," recalls Cooper, "was pointing me straight up the page to 'Imminent Birth,' imminent birth, imminent birth."

"Have you ever delivered a baby?" asked Beth.

"No, but I think we can help you out if there's any problem. You just need to stay as comfortable as possible, okay?"

"That's obviously coming from a man." laughed Beth, and Cooper laughed, too.

Mike drove up and saw Beth banging on the second-story window. She'd heard the car and wanted Mike to hurry up.

I'm coming as fast as I can, Mike thought. How bad does she need this stuff?

Mike raced into the bedroom and hung up the telephone receiver, which he'd noticed was off the hook.

"You just hung up on 9-1-1!" shouted Beth hysterically.

She screamed at Mike, who was baffled, still unaware of the impending delivery.

"She was saying some words that would make a man blush," recalls Mike. "She really was."

Mike dialed 9-1-1 and was reconnected with Bob Cooper.

"I thought, okay, it's brass tacks time around here," recalls Cooper.

Cooper said the ambulance was on its way, then he instructed Mike to get two towels and a shoelace to tie off the umbilical cord. Surprisingly, Beth had a new shoelace on hand because one month earlier she'd asked her doctor what would happen if she delivered while in the car. "Get a shoelace," said the doctor, "just in case."

"I couldn't imagine that I was going to delivery this baby," says Mike. "I just knew 9-1-1 was going to be here on time."

Cooper walked Mike through the instructions printed on his cards.

"It coming! The head's right at the end!" Mike suddenly shouted. "Okay, okay, okay! The head's coming out!"

Beth pushed and the baby came out—with the umbilical cord was wrapped around its neck. Mike delicately removed the cord, but the baby wasn't breathing. Cooper told Mike to tickle its feet.

"There was just this beautiful, most wonderful scream I had ever heard," recalls Beth.

At 7:04, exactly nine minutes after her water had broken, Beth delivered her baby.

"Is it a boy or a girl?" asked Cooper.

"It's a boy!" replied Mike.

"Congratulations," said Cooper.

At that moment, Mike heard the ambulance, and said good-bye to Bob Cooper with great relief.

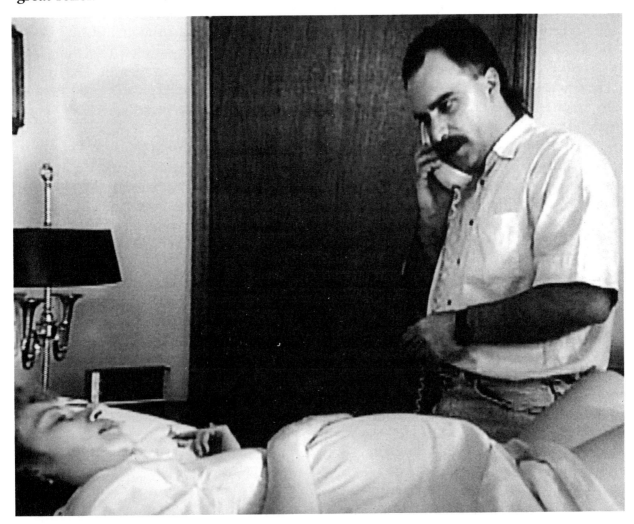

"As soon as I hung up the phone," recalls Cooper, "my head hit the desk and I thought, I can't believe I just survived that. I feel like I gave birth myself."

Beth and baby Drew were examined at the hospital and released in perfect health two days later. Beth thinks her husband was very courageous, and she praises dispatcher Cooper for helping deliver her son.

"I've tried to think of the right words to thank Bob Cooper for all that he did," says Beth. "And I just don't think that they make words that big."

Says Cooper, "The fact that we brought a new life into the world was a pleasant experience—when it was all over."

Collapsed School Bus Driver

Each day, more than fifty children ride to school with Dunellon, Florida, bus driver Donald Snyder. On the morning of December 18, 1991, twelve-year-old Kenny Perrone was one of six boys whom Mr. Snyder picked up at his first stop, at around seven o'clock. Kenny is a special education student with Attention Deficit Disorder, a condition resulting from a lack of the chemical in the brain that allows us to focus on one noise and block out others. Because it is difficult for Kenny to concentrate on one thing at a time, he is usually distracted and inattentive.

Kenny always sat in the rear of the bus, but that morning he took a seat right behind Mr. Snyder. As the bus approached the right turn into Dunellon Middle School, Kenny noticed that Mr. Snyder wasn't looking very good.

"What's the matter, Mr. Snyder?" asked Kenny. "What's the matter?"

Mr. Snyder started gasping for air and couldn't respond. His bus missed the turn to the school and began weaving on the road. Kenny realized that Mr. Snyder had slumped over and lost consciousness.

Kenny jumped out of his seat and got next to Mr. Snyder. Almost simultaneously, the children all noticed they had passed the school's entrance. When they looked at Mr. Snyder and saw him slumped over in his seat, instantly panic broke out. Kenny took charge and yelled to the kids to move to the back of the bus in case it crashed.

Kenny's classmate, Ehren Brown, joined Kenny in an effort to take control of the bus, which was heading for a telephone pole. While Ehren struggled to hold up Mr. Snyder, Kenny tried to peel his foot off the gas pedal and grab hold of the steering wheel. The children screamed in fear, but Kenny managed to steer the bus away from the telephone pole and get his foot on the brake pedal. He pushed hard on the brake and brought the bus to a stop, almost five hundred feet from where Mr. Snyder lost consciousness.

Kenny pulled the door lever and his classmates fled from the bus. One of the students ran to get the assistant principal who was patrolling the school yard, and another notified the school's front office that he thought Mr. Snyder had suffered a heart attack. The office called 9-1-1 and rescue units from Munroe Regional Ambulance Service were dispatched to the scene.

When rescuers arrived, Mr. Snyder was regaining consciousness. He did not complain of pain, but he recalled experiencing stomach cramps right before passing out. Through questioning, paramedic Greg Bicksler suspected that Mr. Snyder's blackout was the result of a severe reaction to prescription cough medicine he had taken that morning for the first time.

As Mr. Snyder was loaded into the ambulance, Kenny wished him well.

"He felt pretty bad," recalls Kenny. "It's scary when you see somebody get sick. He had been a friend to me and I wanted to help him."

At the hospital, tests confirmed the paramedic's suspicion.

"A lot of people take cold medicines," says paramedic Bicksler. "But very few actually think about what they're taking." He advises people to allow themselves at least thirty minutes before engaging in activities that require good judgement, such as driving, after taking medication, including cold medicines. This is done as a precaution in the event they have a negative reaction to the drug.

When Mr. Snyder learned what Kenny had done, he was more than grateful to him. He was also surprised that one of his smallest passengers would take command the way he did. Kenny's classmates and teachers were also a little surprised, but totally impressed by Kenny's bravery and quick thinking.

"What he did was really, really, really heroic," says one student. "I don't think anybody else would have had the guts to do what he did."

Susan and Andy Perrone are extremely proud that their son knew how to handle the emergency.

"Kenny likes to take charge of things," says Andy, a Volunteer Fire Chief. "That day he took charge of things the proper way."

Kenny and Ehren Brown were honored by the city of Dunellon with an award for their heroic efforts in stopping the school bus.

Helicopter Horse

On May 17, 1992, Sherrie Giangrossi, Jack Snider, and another friend, set out for a day of horseback riding in the California Sierras. It was the first time the experienced equestrians were riding this particular trail, and they decided to turn back because it was becoming too steep. Sherrie was doing just that when her horse, Sinbad, lost his footing and plunged down a steep, rocky cliff. Twenty feet below, Sherrie grabbed a concrete pile protruding from the rocks and, miraculously, escaped serious injury. But she had to watch in horror as her beloved Arabian gelding plummeted three hundred feet and landed on a rocky ledge.

Sherrie and her friends scrambled down the ravine and found Sinbad lying on his side. He wasn't moving, and Sherrie thought he was dead. Jack thought that if Sinbad were alive, there was no doubt he would have to be put away. In fact, Sinbad was alive. Jack and Sherrie looked him over and helped him to his feet. He was shaking badly and had numerous cuts, but he didn't appear to have any broken bones.

Sinbad's veterinarian, Dr. Verne Thacker, was called to the canyon. He was amazed that Sinbad was alive considering the great distance the horse had fallen, but he cautioned Sherrie to remain realistic. A trauma such as the one Sinbad suffered could result in numerous complications, including internal injuries.

The pressing question was, how to get Sinbad safely out the canyon? He couldn't walk, and he couldn't be carried because the terrain was too steep and rocky. Dr. Thacker told Sherrie he would try to organize an evacuation plan for the next day. This meant Sinbad would have to spend the night in the canyon, and he would have to spend it standing up, since Dr. Thacker thought Sinbad had probably fractured an elbow, and lying down would increase the problem. Sherrie was not going to leave Sinbad alone, so she and Jack camped out with him.

The next morning, Sherrie's husband, Fred, called around to see how and who might be able to rescue Sinbad. Dr. Thacker thought the most practical way would be to airlift the horse using a special harness designed for large animal rescues. Thacker knew Scott Baker, a helicopter pilot who had prior experience with animal airlifts.

The rescue mission was put into action. Baker arrived on the scene with Dr. John Madigan and his veterinary team from the University of California at Davis. Joining them was Charlie Anderson, co-inventor with Madigan of the special harness. It would be a risky operation.

"Nobody wants to fly in that canyon," says Baker. "There are too many wires and you can't see them. It's windy, there's three bridges, and you've got three hundred fifty feet of cable hanging from the helicopter."

Sinbad was prepared for the airlift. He was blindfolded and sedated so he wouldn't panic, then he was strapped into a sling that would attach to the harness. In a technically difficult maneuver Baker piloted his helicopter between high tension wires and a bridge that obstructed his view of Sinbad. As the helicopter hovered over Sinbad, Baker lowered the harness and the horse was secured. Baker lifted him off the ledge and over the

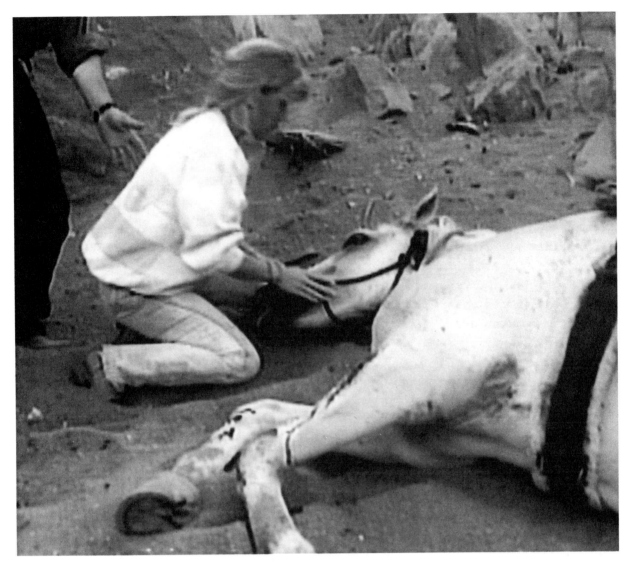

bridge, then flew him five miles to the California Department of Forestry ranger station.

Sherrie was happy to see Sinbad approaching the station, but she knew he wasn't out of danger yet. As Baker lowered Sinbad to the ground, he collapsed. His heart rate was low and Dr. Thacker wondered whether he would lose him on the spot. But once Sinbad received an injection to reverse the effects of sedation, he started moving again. He was transferred to a veterinary hospital where he was treated for a fractured elbow, cuts and abrasions, and bleeding in his chest.

Today, Sinbad is recovering nicely and Sherrie hopes she will be able to ride him in the future. Dr. Thacker is impressed by Sinbad's recovery.

"He had the heart to come through this with flying colors in spite of taking such a bad fall."

Sherrie is grateful to everyone involved in saving her horse.

"Sinbad is family. If it weren't for the rescuers," she says, "I wouldn't have him today."

"Sometimes I wish I could say that I'm my wife's best friend," laughs Jack, "but I think it's Sinbad."

Brain Hemorrhage

On July 14, 1990, New Brunswick, New Jersey, Police Department Dispatcher, Rosa Natal, received two crank calls in a row. They were from youngsters singing, "9-1-1 is a joke." Calls like that make Natal angry.

"You try to call back and explain to the parents that 9-1-1 is not a joke."

A little while later, another emergency call came in. Again, it was a little boy on the line.

"Are you playing on the phone?" Natal asked in a stern voice. She felt an urge to hang up on him.

"No," replied the caller, "I need an ambulance. My mama's sick." Natal heard the panic in the boy's voice and wondered if this call were for real.

"Where's your mom?"

"In the kitchen."

"What's your name?"

"Patrick."

"What floor do you live on?"

"Mama, what floor do we live on?"

When Natal heard a woman call out, "It's a residential home," she became convinced the call was not a prank but legitimate.

EMTs Mike Calabrese and Steve Pawlak were dispatched to the residence from Robert Wood Johnson University Hospital. When they arrived at the house four minutes later, two little boys were standing outside, frantically waving down the ambulance. They were five-year-old Patrick Hammond, with whom Natal had spoken, and his three-year-old brother, Matthew.

The paramedics found the boys' mother, Joan Hammond, seated at the kitchen table. She was pale and vomiting. The medics took her medical history and checked her vital signs, which showed that she had extremely high blood pressure. They were unable to determine the cause of her problem but assessed that she was in critical condition.

A neighbor came over to look after Patrick and Matthew, and as the ambulance pulled away, the boys waved good-bye.

"When the paramedics took my mother away, Matthew was scared," recalls Patrick, "but I told him, 'Our mother's going to be all right.' We both started crying because we really loved her."

A doctor examined Joan in the hospital's emergency room. She was almost unconscious and her condition was deteriorating. The doctor was puzzled about the cause of her medical problem until he ordered a CAT scan, which revealed evidence of bleeding in the brain from a ruptured blood vessel. Surgery was out of the question, because the hemorrhage was too deep within the brain. Joan would have to be treated with medication. For the next forty-eight hours, she hung between life and death.

Happily, Joan responded extremely well to treatment. After two days, she regained consciousness and was able to speak. Three weeks later, she was discharged from the hospital, having suffered no permanent damage as a result of the hemorrhage. When she returned home, Patrick and Matthew were very excited to see her.

"It's overwhelming," says Joan, "the fact that your child has saved your life. I'm well blessed to have these children."

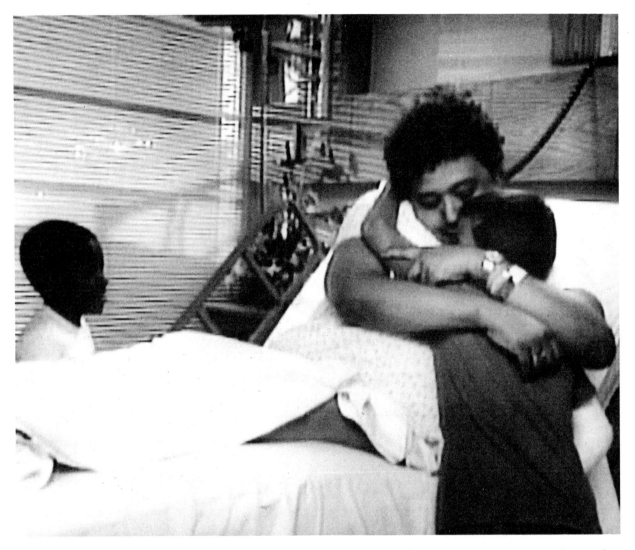

She also praises dispatcher Rosa Natal, who persisted in questioning Patrick so she could determine whether there was a true emergency.

"Patrick's the one who deserves the hugs and kisses," says Natal, "because he did a wonderful job. He's not my son, but I feel proud."

Joan Hammond says parents should explain to their children not to make prank calls to 9-1-1. While they're fooling around on the phone, there may be someone who really does need help and can't get through.

St. Louis Gas Save

On December 19, 1989, Glenn and Annie Murphy and their four children moved just north of St. Louis, Missouri. It was cold inside the Murphys' new house, so Annie asked Glenn to light the furnace while she unpacked boxes. Glenn was aware that the gas company had not yet inspected the furnace, but he decided to turn it on anyway. He then turned on the jets and ignited the pilot. While the three older children, ranging in age from eight to fourteen, were at school, and Annie and six-year-old Queeon spent the day unpacking, Glenn shuttled back and forth from their old house to the new one.

Around four o'clock in the afternoon, Glenn returned to the new house with more boxes and the three children who had been at school. He pulled into the driveway and honked the horn, expecting Annie to come outside and open the garage. When Annie didn't appear, Glenn got out of the car and knocked on the front door. Through the window, he saw Annie slowly walking toward the door. She opened it and told Glenn that her eyes and nose were burning. She felt weak and nauseous, and Queeon and she were both tired and had been sleeping. Glenn responded that Annie was probably just tired from the move and would feel better if she rested.

By six o'clock that evening, Annie complained that she was feeling much worse and was barely able to move. She asked Glenn to take her to the hospital, so Glenn left fourteen-year-old Dominic in charge of the other children and drove Annie to Regional Medical Center.

In the emergency room, Annie had to wait to be treated, but she didn't mind because she'd actually started feeling a little better. Besides, her favorite program, *Rescue 911*, was being broadcast over the television in the waiting room. It just so happened that the victim in this particular segment of the show was suffering from carbon monoxide poisoning and was complaining of symptoms identical to Annie's.

"That's what's happening to me!" Annie excitedly told Glenn.

Suddenly, Annie and Glenn panicked about the children. The furnace was still on and all the windows were closed. Glenn got in the car and sped home, praying for his children's safety but fearful that they might not be alive when he arrived. Glenn pulled into the driveway and ran to the front door. He pounded on the door. Through the window, he saw Dominic pull himself up from the floor by the front window curtains.

"Daddy, something's wrong," cried Dominic as he opened the door.

Glenn rushed inside, opened the windows, and told Dominic to go outside. He ran to check on the girls and found all three in a semiconscious state.

A neighbor's call to 9-1-1 sent rescue units to the scene from Moline Fire Protection District and Christian Hospital Ambulance. The children were transported to St. Louis Children's Hospital, where tests revealed very high levels of carbon monoxide in their bloodstream. They were treated with pure oxygen, and they were released the same day with no permanent effects. Doctors believe that had the children remained in their home for another half-hour, they could have died.

Annie was also treated for carbon monoxide poisoning and released that day, also with no permanent effects.

Annie and Glenn realize how close their children came to perishing and have learned the importance of having a new or unfamiliar furnace inspected by a professional from the gas company. They are both very thankful that *Rescue 911* was on the air that night in the emergency waiting room, because it saved their children's lives.

"I love my wife and kids," says Glenn, "and I thank the Lord that he answered my prayers."

Three Men and a Little Girl

On June 11, 1991, Ruth Martinez sunbathed on the bank of Clear Creek in Golden, Colorado, while her daughters, ten-year-old Camillia and six-year-old Anna, played in the water. When Camillia ventured out to where the water looked clear and gently flowing, she latched on to a big rock.

"You're out too far!" Ruth yelled to Camillia. "Come back over here."

Suddenly, Camillia was pulled under by the current, which was more powerful than it appeared. Ruth jumped into the water and attempted to rescue Camillia, but they were swept downstream together. Anna stood helplessly on the bank and watched them float away, thinking they were going to drown.

Downstream, Ruth and Camillia hit a vortex and Ruth lost hold of her daughter. She watched in horror as Camillia drifted out of sight. Ruth fought her way to the bank and pulled herself out. She ran to a nearby campground and called 9-1-1.

Detectives Glenn Moore and Stan Ross were at their desks at the Golden Police Department when they heard the emergency dispatch. Clear Creek flowed thirty yards behind the station, and since the reported incident had occurred just one mile away, they decided to head out to the river on the off-chance of seeing the victim.

"I'm not a very good swimmer," Ross said to Moore as they approached the creek.

"I'm not either," replied Moore.

"Okay, partner. I'll save you if you save me."

Neither really expected to see Camillia, but as soon as they reached the creek, her body floated toward them, facedown.

Commander Bill Kilpatrick also heard the dispatch and ran to the creek in time to see Ross jump into the river and instantly get swept away like a rag doll by the fierce current. Then Moore jumped in and managed to grab Camillia. He struggled to keep her and himself afloat as they were carried downstream.

"I can't hang on to her!" Ross heard Moore shout. Ross grabbed an overhanging tree branch and pulled himself up on the bank. As Moore floated toward him with Camillia, Ross reached out and latched on to her while Moore grabbed a branch.

Together they pulled Camillia out of the water and checked for a pulse and respiration. Nothing. They would have to start CPR immediately, but first they had to get her body on flat ground. The men slipped and struggled up the steep embankment, joined by Commander Kilpatrick, who had swum across the creek to help. The three officers now carried Camillia uphill.

Ruth met up with Foothills Ambulance at the campground and rode to where Ross and Moore were administering CPR. The officers, emotionally and physically drained when the rescue unit relieved them, wondered what Camillia's quality of life would be were she to regain consciousness after having been deprived of oxygen for so long.

Paramedic Dana Hollingsworth was able to restart Camillia's heart before the ambulance departed for Lutheran Medical Center.

Camillia arrived at the hospital severely hypothermic, with a core body temperature

of less than 84 degrees. Dr. Carla Murphy examined her and did not expect her to survive. Camillia was transferred, in a deep coma, to Children's Hospital in Denver.

Ruth stayed by her daughter's side for three days and nights and prayed for her survival. On the third night, Camillia stirred.

"Come to my mouth," she whispered to Ruth.

Ruth put her ear to Camillia's mouth.

"God," she whispered, "God sent me back."

"Shivers started at the bottom of my feet and went to the top of my head," said Detective Moore when he heard that Camillia had survived. "It's a miracle. I don't see any other explanation for it."

Camillia went through intensive rehabilitation to relearn how to do almost everything, including how to walk and talk. Today, she has completely recovered and leads a healthy, normal life.

Ruth Martinez is extremely thankful to the three men who risked their lives to save her daughter.

"They're angels," says Ruth.

Officers Ross and Moore and Commander Kilpatrick were honored for their heroism by the Golden Police Department. Their greatest reward, however, has been the appreciation shown by the Martinez family.

Anna is also happy her older sister made it.

"I'm glad Camillia is back to her old self because we can do things together again, like make fun of each other."

9-1-1 Trunk-Trapped Tot

On August 31, 1992, Joann Paniccia and her three-year-old son, Paul, of Brookfield, Connecticut, returned home after grocery shopping. Joann and Paul talked as she unloaded grocery bags in the kitchen. After a few minutes, Joann didn't hear a peep out of Paul and realized he wasn't in the room. She called his name as she took a look around the house, but there was no sign of him.

Joann figured Paul might be in the backyard since that's where his toys were, but she didn't find him.

"My heart started beating faster and faster when I heard complete quiet," recalls Joann. "That was scary."

She nervously ran around to the front of the house.

"Paulie, where are you?" Joann yelled. "Paulie!"

She heard Paul's muffled voice respond, "I'm hiding."

The voice was coming from Joann's car, but she remembered having locked it. She looked in through the windows. Paul wasn't inside.

"Paul?" shouted Joann, staring at the trunk.

"I'm hiding, Mommy," came the voice from inside the trunk.

"Mommy will be right back. I'm going to get the keys."

"No, Mommy. I have the keys," said Paul.

Joann began to panic. Paul had locked himself in the trunk of the car along with the keys, and she didn't have a second set. Joann didn't know how long Paul would last without air. She forced herself to remain calm, told Paul she would be right back, then ran inside and called 9-1-1. By the time she hung up she could already hear sirens.

Brookfield Police Sergeant Charles Colman was less than a mile away when he was dispatched to the scene. Joann told Paul that a policeman was trying to get him out.

"You're not going to arrest me, are you?" Paul shouted.

Sergeant Colman tried to unlock the car door with a metal "slim-jim" but his efforts were in vain.

Within moments, rescuers from the Brookfield Volunteer Fire Department arrived with Fire Marshal Wayne Gravius in charge. Gravius talked to Paul through the trunk and assured Joann that he had enough oxygen to breathe. Firefighters assessed the situation and decided they would have to break a window to gain entry to the car.

"What's going on out there?" yelled Paul as he heard the crash of broken glass.

Joann explained to Paul that a firefighter was getting into the car and was going to reach into the trunk through the backseat. The firefighter slipped his hand under the seat and into the trunk.

"Give us the keys, Paul," said Gravius.

"No," responded Paul. "I'll only give the keys to my mommy."

"Paul. Give him the keys," ordered Joann.

Paul reluctantly handed over the keys. Gravius opened the trunk and lifted Paul to safety.

"Why did you go in the trunk, Paul?" asked Gravius.

" 'Cause King Kong told me to go in there," he said.

Gravius tried to impress upon Paul the seriousness of the situation and told him that he could have been in a lot of danger.

Paul simply responded, "You're going to be in trouble for breaking my daddy's window."

When Paul's father, Glen, returned home from work that night, Paul ran to greet him.

"He came running out," says Glen, "and he pointed down at the glass broken in the driveway. He goes, 'Daddy, I did not do it.' "

Gravius also told Paul that he was too young to have his mother's car keys. But, says Joann, tell that to her husband and her father, who get a kick out of giving keys to Paul and encourage his interest in mechanical devices.

"We like to raise manly men," says Glen.

"I'm never going to go in the trunk again," says Paul. "Never. Never more. Never."

Unwanted Intruder

Sheila Parkin of Salt Lake City, Utah, was home alone on the evening of March 18, 1992, while her husband visited his brother a few blocks away. She was upstairs watching television in a back room when the doorbell rang. Since Sheila had earlier turned off all the lights, she walked through the dark to the front of the house. She peered through the window blinds because she wasn't expecting visitors and didn't want to open the door. Sheila didn't see anyone and, feeling uneasy, locked the side door's deadbolt and went back upstairs.

Sheila turned off the light in the television room and resumed watching TV. Suddenly, she heard the sound of her basement window breaking. Sheila turned off the television and dialed 9-1-1.

Dispatcher Crystle Newbold answered her call. In a hushed, frightened tone, Sheila reported that someone had broken into her house and was in the basement. Immediately, units from the Salt Lake County Sheriff's Department were dispatched. As Sheila reported the intruder's movements to Newbold, the dispatcher relayed Sheila's information to sheriff's deputies who were en route to the house.

Newbold was also concerned that Sheila sounded so petrified she was almost hyperventilating. Sheila said she couldn't talk because she heard the intruder coming up the stairs. Taking the phone with her, Sheila crouched on the floor between the couch and the VCR cabinet, pulling a blanket over herself. Newbold continued to ask questions and told Sheila to answer with a yes or a no.

The intruder entered the room and shining his flashlight around, passing it over the blanket and stopping on the VCR, just eight inches away. Sheila was sure she was a dead woman. She was sitting on the VCR's wires, and she could feel them tighten under her as the man tried to remove the appliance from its cabinet. Unable to free the VCR, he left the room. Sheila whispered to Newbold that she could now hear him taking jewelry from her bedroom.

Sheriff's Deputy Paul Barker was the first to arrive on the scene. He parked down the street to avoid being seen and hurried to the house.

"Is there a flagpole in front?" Barker radioed Newbold.

"Is there a flagpole in front?" Newbold asked Sheila.

"Yes," answered Sheila, and Newbold radioed confirmation back to Barker.

Within seconds, Deputy Mark Wooten and Corporal Allen Spencer arrived. It was decided Barker would go around the back of the house and the others would cover the front. Barker was concerned that if the suspect did anything to Sheila, it would take a minute before the deputies could get into the house.

Sheila told Newbold she could hear the intruder leaving through the side door. Newbold radioed to the officers. Barker ran to the side door and stood in position with his gun drawn. When the robber stepped outside with a pillowcase filled with stolen goods, Barker confronted him.

"Sheriff's Department! Freeze!" he shouted.

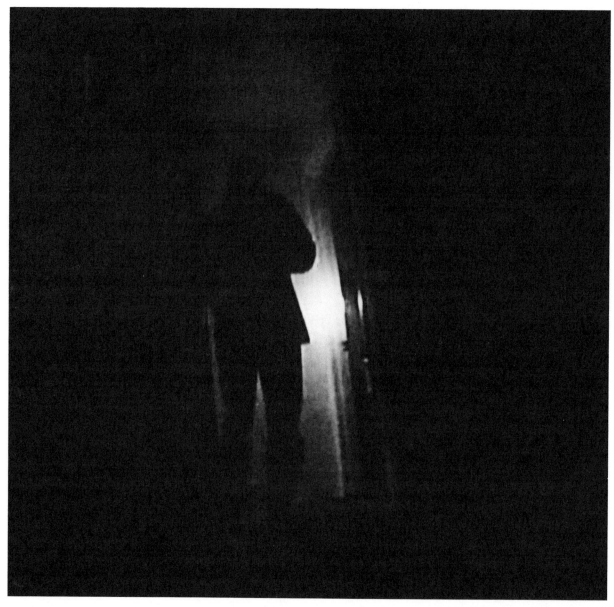

The shocked suspect dropped his pillowcase load and ran back inside, with Barker in pursuit. A moment later, the intruder crashed through the living room picture window and landed facedown in broken glass directly in front of two officers. Barker ran out the front door and handcuffed him.

Officers Spencer and Wooten went inside to check on Sheila. They found her hiding under the blanket, unharmed, and still on the phone with Newbold.

The suspect was taken into custody and found guilty of two felonies. He was sentenced to five years in prison.

Shortly after the robbery, Sheila met Crystle Newbold for the first time. Sheila had wanted to thank her with a hug. It was a happy and emotional encounter for both of them.

"I feel close to her even though I don't know her," says Sheila. "I depended on her for my life. I don't know what I would have done without her."

"Meeting her was like the end," says Crystle tearfully. "Like the closing of a book, and I know she'll go on and do great."

Sister Abduction

Around noon on December 15, 1991, Richard Hagemann gave his daughters permission to play outside their apartment. Six-year-old Shannon and her five-year-old sister Rachel romped on the grass while their father sat in the living room and watched television. Richard kept the front door open and checked on the girls every five to ten minutes.

Shannon and Rachel were playing tag when a young man called to them through a large hole in the fence that separates their yard from their neighbor's carport.

"Look what I found," he said, sticking a handful of dollar bills through the hole. "I know where we can find more. You want to help me look?"

Shannon and Rachel, intrigued by the offer, squeezed through the fence. The stranger cajoled them to go with him to his car, parked nearby. Rachel, who was wary of the situation, lagged behind, but Shannon climbed in the car, coaxed by the man's promise of finding more money and an ice cream. The man told Rachel he'd come back for her and started his engine.

Rachel ran back to the apartment and excitedly told her father that somebody had taken Shannon.

"Where?" asked Richard as he grabbed Rachel and raced to the street.

Rachel pointed to the lime-green compact car driving away. Richard ran back to the apartment and dialed 9-1-1. As he reported the kidnapping to San Diego County Sheriff's Department Dispatcher Don Bronson, Richard feared for Shannon's life. He thought of the recent rash of kidnappings in the area and remembered that not one victim had been found alive.

A "Code Five" was sent over the radio, which means that responding officers would stake out likely areas where the suspect might flee. It is up to the individual deputy to decide his stakeout location, and Deputy Nelson Prosper, one of the responding officers, chose to stake out the Bonsall Bridge, which leads to a rural area.

"I tried to put myself in the suspect's shoes," recalls Prosper. "I thought, if I wanted to get away, which way would I go?"

Deputy Prosper took up his position and within fifteen minutes he observed a small lime-green car. As it came closer, he noticed a child sitting in the passenger seat. Prosper thought this could be his man. His adrenaline started pumping.

"Guys like this," says the deputy, "you don't know what they're capable of—killing the cop, killing the kid, whatever it takes to get away."

Prosper followed the car over the bridge and flashed his lights. The driver pulled over and Prosper approached. He asked the driver what his relationship was to the girl and the man nervously replied she was a friend. Prosper knew right then that this was his man. He ordered the suspect out of the car and handcuffed him.

Prosper comforted the little girl, whom he confirmed was named Shannon, and who was now crying for her daddy.

"It was like I was picking up one of my own," recalls Prosper. "I almost cried, telling her to calm down. You get that big knot in your throat. It was incredible."

The suspect was taken into custody, convicted of kidnapping and child molestation, and sentenced to twenty years in prison.

Richard Hagemann praises police for having done such an excellent job. He thinks it's a miracle that Shannon was found within a half-hour of his call to 9-1-1.

"It's hard to describe how you feel getting a loved one back after a kidnapping," says Richard. "You have your child back and you're going to make sure she's safe for life."

Shannon's mother, Mary, feels very lucky that Shannon was returned. The incident has taught her that she can't be too cautious.

"No matter where you live, or how many times you tell your children not to talk to strangers, they are still children and they're still suseptible," she says.

Rescuing Shannon is one of the highlights of Deputy Prosper's years of service. "As long as I live, I'll remember this part of my law enforcement career for one of the greatest feelings—being a cop and doing something good for a kid. That's what it's all about."

Aussie Whale Rescue

On August 16, 1991, beachgoers watched helplessly as a thirty-foot humpback whale became grounded on a sandbar fifty feet offshore at Peregian Beach in Queensland, Australia. The first rescuer to arrive on the scene was Underwater World's curator and manager, Chris Warner. He thought the situation didn't look good for the whale. The air temperature was hot, the tide was receding, and the whale was burning.

Recalls Warner, "He was in fairly serious danger of boiling in his own blubber unless something was done."

By the time the Sea World marine rescue team arrived, led by coordinator Trevor Long, a large crowd had gathered around the whale. Long was concerned for both the whale and the spectators.

"If the animal moved his tail in a violent action," says Long, "you've got four tons of body mass that could break you or even kill you."

Long and his team devised a plan to free the whale by attaching it to a sling—which Long would have to design—then tow the whale out to sea with *Sea World One*, a sixty-foot rescue and research vessel. But towing would have to wait until the next high tide, which would occur at midday the following day.

Throughout the night, volunteers rotated in thirty-minute shifts, braving cold air and water temperatures, to keep the whale facing into the waves so it wouldn't be hit broadside and drowned.

Meanwhile, Long and Warner devised a sling that would be self-releasing so that if the whale went off on its own while being towed, it wouldn't sink *Sea World One*. Neither man had any experience in moving such an enormous animal off the beach, or in designing such a sling.

"It's in the middle of the night," recalls Long, "we're all tired, we've been there all day. We're wet, we're cold. There's no book that you can look up and say, whale sling, thirty tons, what am I going to use?"

By the next morning, the crowd on the beach had swelled to thousands. Long and Warner arrived with their sling and were met with a depressing sight. The tide had receded further and the whale was surrounded by sand. Its breathing was labored, and it seemed to Long as if the animal had given up its will to live.

But the volunteers were not about to give up on the whale. It was now a race against time as they dug with shovels around the whale to build a sand wall. The wall would stimulate wave action.

"It was amazing to see thousands of people helping that one whale," says Long, "and to think that only a few years ago we were hunting these whales. It was quite heartening."

Still, Long knew the situation was desperate because of the receding tides and the whale's worsening condition. The high tide would be only one window of opportunity to save the whale. As Long coordinated efforts to dig a channel from the whale out to deeper waters to position it for towing, a man approached and asked what he could do to help.

"Can you dig me a channel through the surf?" Long jokingly asked.

"I'll give it a good go," he replied.

Moments later, the man reappeared in his earth-mover, drove it into the surf, and started digging a channel with his heavy piece of equipment.

As the tide rose, the surf became unexpectedly rough, jeopardizing the rescuers' safety and the entire rescue operation. But the whale remained calm. Long believes it knew they were trying to help it. Now the challenge became slipping the sling's loops over the whale's fins.

"Putting loops over pectoral fins," recalls Long, "is like putting on a tight pair of jeans while standing in a crowded, moving bus."

While *Sea World One* rocked and rolled in the rough surf, the sling was attached to the whale. Rescuers signaled for the boat to start up its engine. It was an emotional moment for volunteers and the five thousand cheering spectators as the whale was towed off the sandbar.

"I had tears in my eyes," recalls Long, "and part of me was saying, Jeez, you can't stand here and cry in front of all these people. Actually, I felt like going down on my knees."

The minute the whale got moving, it swam free of its sling, passed the boat, and disappeared into the ocean. It was the first time that a beached animal of this size had ever been successfully rescued.

Heart Attack Crash

On July 20, 1989, in South Bend, Indiana, Faith United Methodist Church secretary Mary Ferrel was talking on the phone in the church office. Sixty-four-year-old Robert Carl, a Bible Studies teacher, walked in looking pale and sweating. He complained of chest pains and asked Mary to call his son, Dale, who was a paramedic.

Dale was attending a class at a hospital across town when he received news of his father's condition. He grabbed his partner, EMT Randy Garnett, and hopped into an ambulance.

"I was worried Dale wouldn't be able to handle the call if it was bad," recalls Randy as they sped to the church, "because the worst possible scenario is responding to someone you know."

When Dale saw his father, he immediately knew he was having a heart attack. Robert looked ashen, was short of breath, sweating profusely, and was in so much pain he couldn't speak. Dale gave his father nitroglycerin to ease the pain, but it had no effect. He was in a dilemma. He might need to perform CPR, but his father had earlier told him not to restart his heart if it ever stopped.

"But you just don't want to lose your best friend," says Dale.

As Randy raced to the hospital, he slowed the ambulance as he approached an intersection. It appeared safe, so he proceeded, but just then a car sped through the intersection and crashed into the ambulance, causing it to roll over three times and land on its side.

The ambulance was in a shambles. Heavy equipment had been thrown all over. Worse, the force of the crash had thrown Robert's gurney upright and slammed him into the cabinets on the wall. He was suspended in air from the gurney's safety straps and his IV line had broken loose from the catheter still in his arm. His head was cut and bleeding.

Dale called out to Randy, who was pinned between the door and the seat. Randy waved his arm in the air to signal that he was okay. He grabbed the radio and informed the hospital of the accident.

Dale's sister, Terry Maguire, and his wife were awaiting Robert's arrival at the hospital when they heard news of the crash.

"I thought, I've lost my father and my brother in the same wreck," recalls Terry.

Dale's left arm felt as though it were broken, but he crawled to his father and tried to stabilize him. He found the lost IV, cleaned it, and plugged it back in the catheter. He lowered Robert and put his oxygen mask back on. As his father's condition deteriorated, all Dale could do was to comfort him and wait for help to arrive.

Rescue units from the South Bend Fire Department arrived on the scene within four minutes. Rescuers extricated Randy from the wreck, while paramedics loaded Robert into another ambulance and prepared to leave without Dale, who was to be transported in still another ambulance. But Dale insisted on accompanying his father.

"I started care and I'm finishing care," he told paramedics. "This is my dad and I'm going with him."

At the hospital, Randy and Dale were treated for their injuries. Robert was admitted to South Bend Memorial Hospital in critical condition. Tests revealed blockages in four major arteries, and the following morning, Robert underwent a quadruple bypass. Dale spoke to his father as he was being prepped for surgery.

"You better make it," Dale told him, "because I'm not ready to be the man of the family."

Robert assured him he would. His surgery was successful, and he was released from the hospital one week later. Today, Robert Carl is doing very well. He appreciates every minute of the day because he knows he's living on borrowed time.

"I thank God that we can joke about it now," says Dale, "and say that we're just two old country boys that took a whooping."

Dale's wife thinks that motorists should learn a lesson from the accident.

"The most important thing people need to remember is, if you hear a siren, pull over."

Sporting Goods Heist

December 21, 1991, was an unusually slow Saturday morning for M & S Sporting Goods in Mesa, Arizona. The owners, sixty-nine-year-old George Rosenkrans and his wife, Emilie, were working in the back along with their daughter, Judy, and employee, Rick Meyer. There were no customers in the store when George heard the doorbell chime and walked toward the front of the store.

"May I help you?" he asked.

George was greeted by three masked gunmen.

"This is a holdup!" announced the thief disguised in a gorilla mask as he aimed his sawed-off shotgun at George.

A second gunman, wearing a ski mask, shoved his shotgun into George's stomach.

"You've got to be kidding," the stunned owner replied, pushing the gun aside. The gorilla-masked assailant responded by shooting George in the side.

Judy heard the blast and immediately knew her father had been shot.

"Dad!" she screamed.

Instantly, another shot rang out and a bullet whizzed past Judy's face, missing her by inches. She hid under a desk and wondered whether her father were dead or alive. The gunmen then ordered Rick to the front of the store and forced him to fill their bags with pistols and rifles. Then they ordered him to lay facedown on the floor. Rick expected a bullet in the back. Instead he heard the door's automatic chime ring as the gunmen fled.

Bob Ramsey happened to be driving past the store when the masked men ran out, their arms loaded with rifles. He knew immediately they had just committed a robbery and, since he wanted to see them caught, decided to follow them.

Meanwhile, in the store, Emilie called 9-1-1. Her emergency call was answered by Mesa Police Department Dispatcher Gary Melton, who sent police and rescue units to the scene.

Ramsey followed the assailants to an apartment house. He was concerned they might see him, but the men were too busy unloading their rifles to notice. Ramsey then drove back to the sporting goods store to phone the police, but by the time he arrived, officers were already on the scene, including investigating officer Lieutenant Dan Day. Ramsey told Day he knew where the suspects were.

While Mesa Fire Department Paramedic Arnold Cornejo treated George, rushing him to the hospital, Lt. Day drove Ramsey to the suspects' hideout. Backup units arrived as Day and Ramsey watched for some sign of the suspects. Officers sneaked up on the apartment. When the door opened and the men walked out, Ramsey confirmed they were in fact the suspects and the police made their move, apprehending all three. The suspects were later convicted of armed robbery and aggravated assault and sentenced to prison.

George underwent surgery to repair his gunshot wound and remove a damaged spleen. During his recovery he suffered a stroke, but after intensive physical therapy and a difficult recuperation, he returned to work seven months later. For safety, he's stopped selling guns and gun accessories in the store.

"I went through five campaigns in World War I and II never got hit. Civilian life is rough," laughs George.

Lieutenant Day was very impressed with Bob Ramsey's actions.

"What Mr. Ramsey did is far beyond what's expected and what's asked. Thank God he wasn't injured. But," adds the lieutenant, "I wouldn't recommend that anybody take that much action without the proper training or ability to protect yourself. The potential for violence in this situation was extremely high."

George and his family are thankful that he's okay.

"I value each and every day that I get up and go at it again," says George. "My wife and I have been married forty-five years and I tell her I love her every single night now."

"I love George," says Emilie. "I don't know what I'd do without him."

Cribbage Choke

On the evening of April 12, 1992, four-year-old Shane Luther and his two-year-old brother, Kyle, of Honeoye Falls, New York, were playing in the living room. Their mother, Kay, was cleaning nearby.

"Try this, Kyle," Kay heard Shane say.

Shane put a cribbage game piece, which is shaped like a golf tee, into his mouth and spit it out like a rocket ship.

A moment later, Kay heard a gasping sound. She turned around and saw Shane clutching his throat with his hands.

Kay yelled to her husband, John, that Shane was choking. John ran into the living room and slapped Shane on his back, but the game piece did not dislodge. John had not been trained in the Heimlich Maneuver, but he attempted the technique based on what he'd seen on television. When he couldn't expel the object, John tried to look down Shane's throat, but his son's jaw was rigidly clenched.

Since the Luthers' area was not equipped with the 9-1-1 system, Kay dialed the local emergency number. Her call was answered by dispatcher Murray Henry of the Ontario County Sheriff's Department. Kay stayed on the line with Henry while the Honeoye-Richmond Volunteer Rescue Squad and an Ontario County Advanced Life-Support Unit were dispatched to their home.

Henry instructed Kay through the step-by-step details of how to perform the Heimlich Maneuver. Kay relayed the instructions to John, but John still couldn't dislodge the cribbage piece. Shane was losing consciousness.

The Luthers' neighbors, local Assistant Fire Chief John Mason and his wife, Sandy, heard the emergency dispatch over their radio and rushed to the house. Mason was also unsuccessful in expelling the game piece. Shane was not getting air and his condition was deteriorating rapidly.

When the Honeoye Falls-Richmond EMTs arrived and loaded Shane into the rescue vehicle, Mason felt a sense of panic and helplessness. It was the first time in twenty-five years of service he had felt this way, and he was thinking that this rescue attempt would probably end in tragedy.

The situation was too critical to wait for advanced life support, so EMTs radioed the paramedics, who had been dispatched from F. F. Thompson Hospital, twenty-five miles away, and instructed them to take the same route, so they would meet up mid-way.

En route, Shane began turning blue and was barely breathing. Further attempts at the Heimlich Maneuver had failed.

Ten miles from the hospital, the ambulance met up with the rescue squad. When paramedic Bernie Leavitt boarded the rescue vehicle, Shane, who had been choking for over fourteen minutes, appeared lifeless. He was not breathing and his heart rate was dropping. Leavitt tried to intubate Shane but was unable to get air into his lungs.

As John Mason drove Kay to the hospital, they listened to broadcasts from the rescue vehicle on John's radio. When they heard that Shane was going into respiratory arrest, Mason turned off the radio.

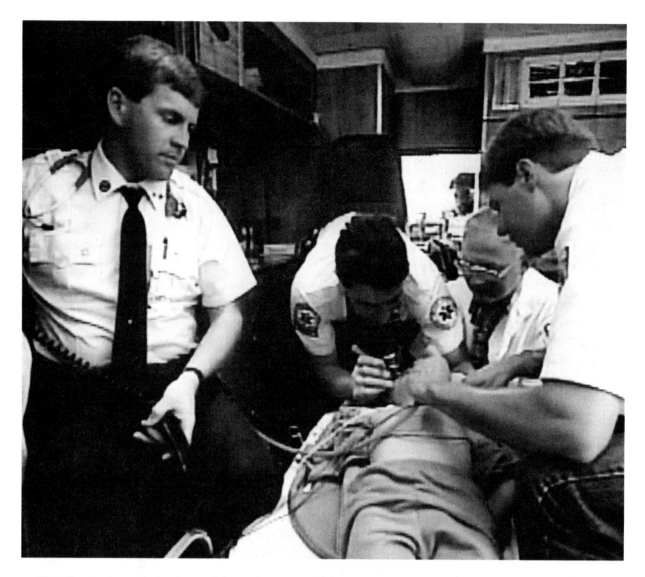

"Without Bernie," says John, "there wouldn't be a Shane."

Shane's experience has made him a wiser and more cautious child. He watches his younger brother carefully and makes sure nothing hazardous goes into his mouth.

"You don't need to hear this," he said to Kay, aware that in another sixty seconds Shane would be in cardiac arrest.

Bernie Leavitt knew he had to act quickly to save Shane. He decided to perform a needle cricothyroidotomy, which he had never done before. He inserted an intravenous catheter needle into Shane's neck, bypassing the obstruction, and pushed air into his lungs. Shane responded with an immediate turnaround—his heart rate increased.

At the hospital, Dr. Tom Benzoni surgically removed the cribbage piece from deep inside Shane's throat. Kay could hear Shane in the emergency room.

"Shane started screaming at the top of his lungs," she recalls, "and it was music to my ears. That's when I knew he was all right."

Shane was transferred to Strong Memorial Hospital in Rochester and released four days later without any permanent damage.

Kay and John Luther are grateful to all the rescuers for saving Shane's life, especially Bernie Leavitt.

Heimlich Brothers

In many parts of the country, the Camp Fire organization goes into schools to teach children a self-reliance course called "I Can Do It." The idea behind the program is that no matter how young children may be, they can learn to take care of themselves.

On April 24, 1989, Donna McEnany of the Camp Fire Organization was conducting a chapter of the course for second-graders at Runnels School in Baton Rouge, Louisiana. The topic being covered that day was "I Can Handle a Real Emergency." Seven-year-old Kyle Wilkinson volunteered to role-play a choking situation with Donna in front of the class. Donna then demonstrated the Heimlich Maneuver on Kyle, instructing students when and how to administer the proper abdominal thrusts on a victim.

Caryl Wilkinson picked up her son after school. Kyle was riding in the backseat of the van with his brother and sister, three-year-old twins Joshua and Jessica. Kyle and his mother chatted about his schoolday on the drive home. He was telling her how his classmate had thrown worms on their teacher when suddenly he noticed what Joshua was doing.

"Mom, Joshua's choking! Stop the car!" yelled Kyle.

Caryl looked in the rearview mirror and saw that Joshua was purple. She stopped the car and jumped out, and by the time she ran around to the passenger side. Kyle had already pulled Joshua out of his car seat and was standing behind him on the side of the road, his arms wrapped around Joshua's small body as he attempted to do the Heimlich Maneuver.

"Give him to me!" Caryl screamed at Kyle. She didn't know why Kyle was hugging Joshua like that.

"No, Mama. I'm doing the right thing!" cried Kyle as he administered abdominal thrusts.

"You're hurting him! Give him to me!"

Kyle ignored his mother and continued administering thrusts. Suddenly a piece of candy popped out of Joshua's mouth. Joshua started crying, and Caryl knew he was breathing on his own.

Within three hours of being taught the Heimlich Maneuver, Kyle had used it to save his brother's life. "What's even more amazing," says his second-grade teacher, "is that Kyle took the situation into his own hands when his mother didn't know what to do."

"I was surprised because everything worked out well," says Kyle. "It makes me feel glad because Joshua's still alive and I still get to play with him."

"I'm really proud of Kyle," says Caryl. "Not just because he saved Joshua's life, but because he's shown other people and other children that you can do something like this, you can help."

Caryl praises the Camp Fire Organization for their work.

"I was so grateful that they taught Kyle the Heimlich Maneuver. They'd done something I really needed, something everybody needs to learn."

Mountain Pool Plunge

On August 26, 1991, David Salinas and Jon Kubar hiked to California's Angel Falls, a series of waterfalls that flow over rock formations and empty into many different pools. The twenty-three-year-olds planned to spend the afternoon swimming and sliding on the falls. They didn't see the posted Danger sign warning of death to those who have attempted to slide down the rock formations into the waterholes below.

Jay Rassi and his sons, ten-year-old Brandon and eight-year-old Adam, happened to be hiking down the mountain trail across the canyon from David and Jon. Jay's sons stopped to watch the young men. Jon was poised to slide down a cascading stream on the slick rock into a pool below.

"Let's go," said Jay to Brandon and Adam. Jay, who had grown up in the area and was familiar with the treacherous falls, did not want his sons to see what was about to happen. From his vantage point, Jay could tell that Jon was not going to land where he thought he would, in a pool a short distance below. Rather, he was going to slide straight down the steep rock face and possibly land in a lower rock-filled pool. Jay resumed hiking down the trail, but quickly noticed his sons were not behind him. He backtracked to Adam and Brandon, who were still watching the men. He was just in time to see Jon start to slide, and Jon did exactly what Jay predicted. He missed the upper pools and fell—out of sight.

David had watched Jon take off, but he couldn't see where his friend had landed. Jay, scrambling across the rocks, yelled to David that Jon had fallen as he climbed down to look for him. David, concerned about finding Jon, slid down the same path his friend had taken. He landed in the pool where Jon was—trapped headfirst underwater—but not without hurting his leg on the wall of rock on his way down.

David reached around in the murky water and found Jon's head wedged between two rocks. After three tries, David yanked Jon free and pulled his blue, lifeless body to the surface. Jay thought he was moving a dead body as he helped lay Jon on the rock. David didn't know correct CPR procedure, but he knew he had to help his friend. He reached into his memory and called up some CPR training he'd once taken in high school. After a minute, Jon started breathing and regained consciousness.

David didn't realize how badly he'd injured his leg until he tried to stand up and couldn't. He would have to stay with Jon while Jay and his sons hiked out to summon help. As soon as Jay got to his van, he encountered a Mariposa Park Ranger, who relayed word to Madera County Sheriff's Department dispatch. Rescuers from the Madera County Fire Department and Sierra Ambulance headed to the scene. They arrived at the top of the falls within twenty minutes and evacuated Jon and David from the gorge.

While hospitalized, Jon was treated for head lacerations and a concussion, and had water pumped out of his lungs. David underwent reconstructive surgery to repair a shattered knee and multiple fractures to his left leg.

Jon is alive today, thanks to David and the one CPR class he took in high school, which made the difference in his friend's survival. After the accident, David became certified in CPR because, as he says, "You never know when you're going to need it."

"David is an angel that was sent here," says Jon's father. "To totally forsake his own life to save someone else, he's a true friend."

Jon is also appreciative for David's heroic efforts.

"I've slowed down a little," says Jon. "The accident has made me appreciate the simpler things in life. And I can't say enough about David. I know he's gone through some pain and probably still does, but I'm still around and he's responsible for that."

Sixth-Grade CPR Save

On September 14, 1992, the new fall term at Tracyton Elementary School in Bremerton, Washington, was into its second week, and eleven-year-old Aaron Rankin and his best friend, Chris Sander, were among the students in Randy Corbett's new sixth-grade class. That afternoon, forty-year-old Mr. Corbett, an eighteen-year-veteran teacher, had just begun his math lesson for the class. He was standing next to the table where Aaron and three other students, children he barely knew, were seated. Mr. Corbett was in the middle of writing a math problem on the overhead projector when he suddenly collapsed face-first on Aaron's desk.

Aaron's classmates started giggling. They thought Mr. Corbett was fooling around, but Aaron knew this was no joke. He laid Mr. Corbett on the floor and checked for a pulse and breath. Mr. Corbett had neither, so Aaron, who had been trained in CPR by his parents, began the procedure on his teacher.

There were more chuckles from students when they saw Aaron giving Mr. Corbett mouth-to-mouth resuscitation. But the mood quickly changed when Aaron told Chris to get help.

Chris ran to the main office and told the secretary what had happened. As she called 9-1-1, the school's principal, Ann Lawrie, who had taken annual CPR training for the past twelve years, ran to Mr. Corbett's classroom.

Teacher Ben Pedersen heard the commotion from next door and went into Mr. Corbett's room too. Mr. Pedersen checked Mr. Corbett, found him to be without a pulse or breath, and took over CPR from Aaron. Ann Lawrie arrived, and she and Mr. Pedersen performed CPR together until a gym teacher, who was even more experienced in CPR than Mr. Pedersen, took over and continued administering CPR with Ann. Aaron was in tears as he stood with the other students and watched.

"Aaron was crying and shaking because he thought Mr. Corbett was going to die," recalls Chris. "And he thought he could have done more."

Within six minutes of the emergency call, the Kitsap County Fire District Rescue Squad arrived with paramedic Brian Danskin. As Danskin ran toward the classroom, he noticed a sign pasted on the window announcing "Mr. Corbett." Danskin was struck because Mr. Corbett had been his sixth-grade teacher. It wasn't until he kneeled on the floor to assist with the ventilations that Danskin discovered that his victim was Mr. Corbett himself.

Paramedics shocked Mr. Corbett's heart using a defibrillator, but it did not regain its regular rhythm. They shocked him a second, third, and fourth time without success.

"The time that the paramedics were working on Randy were the scariest and most frightening, most horrifying seventeen minutes of my life," says principal Lawrie. "I wanted the paramedics to be done and I wanted the happy ending to be right then and there."

After the fifth shock, Mr. Corbett's heart reestablished a regular rhythm and he was stabilized enough to be transported to the hospital.

Doctors determined that Mr. Corbett was suffering from an inherited condition that causes the electrical currents in his heart to malfunction, which led to his cardiac arrest. A 'defibrillator' was implanted in his chest to control the problem.

Mr. Corbett was released after seven days in the hospital. Six weeks later, he was well enough to return to his classroom. His wife says Randy is a lucky man. Because Randy was the only member of his family who was trained in CPR, he probably would have died had he experienced his cardiac arrest while at home. As a result, everyone in Mr. Corbett's family has now taken CPR training. Administrators at Tracyton Elementary School were also sufficiently impressed by the incident to hold CPR training for their students.

"I believe that all children should be taught CPR all over the United States in school," says Randy Corbett. "And if they're taught, there will be a lot of lives saved."

The Central Kitsap School District and the local Heart Association honored Aaron Rankin for his heroism and for saving Mr. Corbett's life.

"I was proud of myself and I was glad that I was there that day," says Aaron. "Before it happened, I didn't know what I wanted to do. I'm going to be a paramedic."

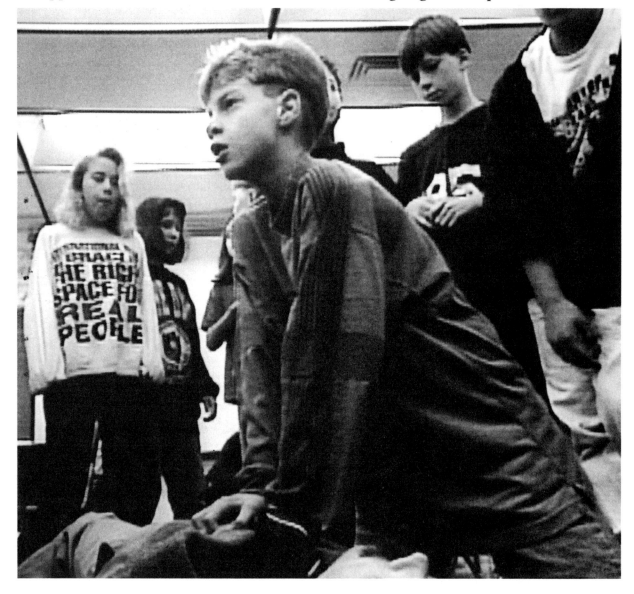

Lady's Not a Tramp

On September 7, 1992, Laura Martinez gathered with family and friends for a Labor Day barbecue in a remote area near Loveland, Colorado. Also enjoying the day was the Martinez's beloved dog, a mixed retriever and husky named Lady.

"Lady is a lovable, gentle dog, but she's very lazy," says Laura's mother, Oma Thomson. "Because she's overweight we always make her walk with us, and by the time we got to the barbecue pit, she was out."

One of Laura's young daughters, Audrey, announced she had to go to the bathroom, and the other children decided they had to go, too. Oma offered to escort the children to a clump of nearby trees and shrubs designated as the latrine. She led her grand-daughters to one side of the creek, while Laura's husband, Bob, led his son in the opposite direction.

Laura's eight-year-old daughter, Teresa, ran ahead of the others to the trees and disappeared out of sight. Suddenly, Oma heard an eerie rattling sound and stopped in her tracks.

"Teresa, stand still!" yelled Oma. "I hear a snake."

Teresa froze. She looked down and saw a rattlesnake at her feet, poised to strike.

"Please, Oma!" shouted Teresa. "Help me!"

"Don't move, honey. I'm coming to get you," replied Oma. Then she turned toward the barbecue pit. "Come quick!" she yelled to the rest of her family. "There's rattlers all around us!"

Oma didn't see them, but she knew from the sound that they had walked into a den of snakes.

Suddenly, Lady leaped to her feet and raced to Teresa's side. She snapped and lunged at the rattler, while Oma instructed Teresa to move back slowly, go around the trees, and run to her.

Oma ran with Teresa back to the barbecue while Lady battled the snakes. Once everyone had run to safety, Laura and Bob screamed to Lady to come, but Lady continued to fight. A moment later she stopped, looked around, and noticed that everyone had fled the area. She sluggishly walked back, and when she reached Laura, dropped to the ground.

Oma, Bob, and Laura examined Lady since Oma thought she had seen the rattler strike her during the fight. Noticing puncture wounds on Lady's face, they loaded her into the car and drove her to Colorado State University Veterinary Teaching Hospital. The children cried as Lady rode away.

By the time Lady arrived at the hospital one hour later, her face had swollen by half its normal size. Veterinarian Lisa Metelman examined Lady and found several snakebites. She told Laura and Oma that their dog had a 50 percent chance of surviving, and she offered the option of treating Lady with a very expensive antidote, an anti-venin serum. Laura didn't need to deliberate.

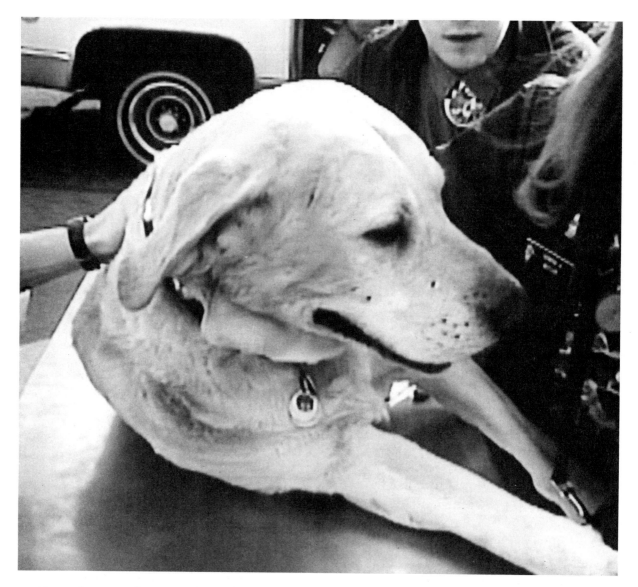

"Do what you have to do," she said, crying. "Lady saved my little girl's life. She took the snakebite for her. She's a hero."

When Laura and Oma returned home, they had an emotional and tearful discussion with the children about Lady's condition.

"Lady might die," Laura told her children. "If she does, we know that she died giving her life for Teresa. She's a very special dog."

That night, the whole family cried themselves to sleep.

The next morning, Laura called the hospital for an update on Lady's condition and was excited to hear that she had responded well to the anti-venin serum. In fact, she was doing so well she would be released that day.

Lady received a warm welcome home from her family.

"It's not a cliché that dogs are man's best friend," says Laura. "Lady is certainly ours." Teresa agrees.

"Lady's a hero. She saved my life. She's the best dog in the whole wide world."

"Today, she's just Lady again," says Oma. "She's great. She's still lazy and she still just sleeps most of the time."

Hanging Hang Glider

July 30, 1989 was a perfect day for hang gliding near Winston-Salem, North Carolina. Experienced hang glider Gilbert Aldrich, and his fiancée, Nancy Lumlee, met two other hang gliders, Doug Rice and Jim Bruton, atop Sauratown Mountain. Nancy, a novice hang glider, helped Gil with his preflight routine. After they carefully checked out his gear, Gil moved into position on the launch site. Because he was strapped into his harness, Gil couldn't reach his helmet, which was hooked to the glider's cross bar. So he unhooked himself from the glider and leaned over to get it. He put the helmet on and felt a good breeze come up.

Gil hurried to take off. Nancy watched Gil fly off the mountain and she immediately knew something was wrong. Gil was heading straight toward the dense forest a half-mile below. Nancy watched helplessly as Gil crashed into the trees. She screamed to Doug that Gil was in trouble.

Nancy and Doug drove down the mountain separately to find Gil, and Jim, who was in flight, glided toward the area to meet them.

'This can't happen to us,' Nancy thought as she drove, tears streaming down her face. They were going to be married.

Jim and Doug met up and hiked toward the accident site. When they approached, Doug saw Gil's glider stuck in the treetops and Gil lying on the ground, badly injured. He'd been in that state for forty-five minutes and had lost a lot of blood. It was the worst hang gliding accident the experienced glider had ever seen.

"I wasn't hooked in," Gil whispered hoarsely to his friends.

The men realized that Gil had descended out of control at a speed of sixty miles per hour, as opposed to the usual twenty-five miles per hour of normal gliding.

Doug ran to the nearest house and called 9-1-1. Dispatched to the scene were Sauratown Volunteer Fire Department EMTs, and Stokes County EMS Paramedics Greg Collins and Thomas Gordon. When the rescuers arrived, paramedic Collins examined Gil and knew he was in serious trouble. Gil's blood pressure was dropping dramatically, he looked shocky, and he was very cold despite the 85-degree air temperature.

It took almost forty minutes to carry Gil to the base of the mountain, where an Air Care helicopter was waiting to transport him to Baptist Hospital in Winston-Salem. Gil was barely clinging to life when he arrived at the hospital's trauma center. Dr. Wayne Meredith determined that Gil had lost half his blood internally and externally and needed immediate surgery to repair a collapsed lung and perforated kidney. He also was suffering from a broken elbow and head contusions.

Gil had a difficult one-year recovery, but Nancy never left his side. During his recuperation, Gil hung up his glider permanently. Now he feels he enjoys life a lot more without the sport that once completely consumed him.

Almost two years after the accident, Gil and Nancy got married. They are deeply appreciative of Doug and Jim and all the rescuers who saved Gil's life.

"You wait all your life for someone and then to think you might lose him," says Nancy. "I was not going to lose Gil."

"That was a long, hard road back to recovery," says Gil, "and Nancy was willing to give everything she had for me. I had never felt that kind of love and care, and I just don't ever want to be without a woman who loved me as much as she loved me then."

1033 Officer Down

In the early evening of July 6, 1991, Utah County Sheriff's Deputy Mike Morgan was on patrol when he saw a man being pursued by two others. Morgan stopped his truck and radioed for backup to Provo Police Officer Phil Webber, who he knew was in the area.

While Morgan physically separated two of the belligerent men, the third man ran away. Officer Webber pulled up, and as he ran to assist Morgan, he noticed a man 150 feet down a dirt lane, approaching with an aimed rifle. Morgan recognized the gunman as the man who had run away.

The officers pulled the two men behind Morgan's vehicle for cover and ordered the gunman to drop the rifle. Instead, he fearlessly continued approaching without cover and opened fire.

"Ten-thirty-three," reported Webber urgently into his hand-held radio, stating his location.

The code "ten-thirty-three" means officer needs help now, so when Provo Police dispatchers Linda Hargadon and Susan Lester heard Webber's voice, they knew he was in serious danger. The dispatchers called for backup and fire and rescue units to respond to the scene.

The officers returned the gunman's fire, but their firearms were not very effective at such a distance. Several rounds were exchanged before the suspect went down. Then Officer Webber stumbled to the back of the truck, blood squirting from his neck.

"I'm hit. I'm hit," he radioed to dispatchers.

"When I heard that Phil was shot, I couldn't believe it," recalls Hargadon. "And then I remembered back to the night before when I called him on the air and told him he'd just become a grandfather again. I kept thinking, this isn't fair."

Morgan figured Webber's artery had been hit. He sat Webber upright against the truck's bumper and pressed his hand on the wound to stop the bleeding. The suspect struggled to sit up and aimed his rifle. Morgan reached for his M-14 assault rifle in the rear of his truck. He couldn't remove his hand from Webber's wound, but he couldn't load the rifle with just one hand. One of the men offered his assistance, and Morgan felt he had no choice but to entrust him with Webber.

Morgan aimed, and the other man jumped him from behind, yelling not to shoot his friend. Morgan commanded the man to get back, prepared to shoot him if necessary. Meanwhile, the suspect with the rifle was not giving up. Morgan found himself in an outrageously dangerous situation, his attention divided between a suspect to his front, one to his back, and a third handling the critically wounded Webber.

Morgan fired several shots and the suspect went down.

"I'd never shot anybody before," recalls the deputy. "It's tough to take a high-powered rifle and shoot another human being."

Before rescuers safely entered the scene, Morgan and Park Ranger Mike Forshee, who had just arrived, secured the rifle from the suspect, who was still alive. Morgan and the

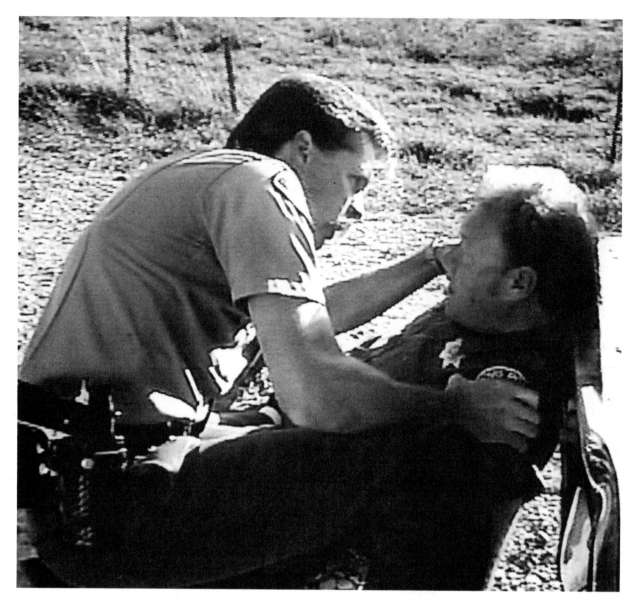

rescuers didn't hold much hope for Webber's survival as he was transported to Utah Valley Regional Medical Center.

At the hospital, tests revealed that, contrary to initial impressions, Webber did not receive any lacerations to his artery, so the shrapnel was left in his body. Doctors say that had the bullet been one-half inch closer to his artery, he would not have survived. Doctors desperately attempted to save Webber's right index finger, which had been shattered, but ultimately they could not.

Says Morgan, "Your trigger finger is pretty much your life. If you don't have a trigger finger you're not a cop."

Today, however, Phil Webber is still a cop, the only Provo police officer qualified to shoot with both hands.

The suspect survived, was found guilty of attempted murder, and was sentenced to fifteen years in prison.

"Phil's a very lucky guy," says his wife, Karen, "and I'm lucky, too, to have him with me."

Officer Webber feels deep gratitude toward Morgan for saving his life. The men, previously professional acquaintances, have formed a stronger bond because of surviving the incident together.

"My wife and I both firmly believe my life was saved by an angel," says Webber. "That took the form of first aid administered by Deputy Morgan, by medical personnel and other officers, and hospital personnel—everything put together. I owe a lot to a lot of people."

Florida Keys Scuba Save

"I guess you could say we were childhood sweethearts," says Walter Benjamin. "I first slept with her when I was six years old, when my little brother was born. My folks took me over and dumped me in her bed. So I had to marry her."

After nearly fifty years of marriage, Walter Benjamin was even more in love with his wife, Joanne, than when he first shared her bed years ago. Besides their love for each other, the couple also shared a love for the ocean. On September 7, 1987, as they often did, sixty-eight-year-old Walter and sixty-nine-year-old Joanne planned to spend Labor Day weekend fishing and scuba diving in the Florida Keys.

Before going into the water Walter and Joanne put on their diving gear and checked each other's tanks. Their custom was for each to fall backward into the water off opposite sides of their boat, then swim to the mooring line in front and dive together.

When Walter arrived at their rendezvous, he found Joanne laying facedown in the water with her snorkel in her mouth instead of her regulator. Walter swam over to Joanne and handed her the regulator.

"I can't breathe," she said. Then Joanne passed out. Walter was scared to death as he swam back to the boat with Joanne in his arms. She was absolutely limp, and Walter didn't know how he was going to get her back on deck.

"Getting around the boat was easy," he recalls. "Getting her into the boat was next to impossible."

Walter yelled for help. His cries were heard by Lauri MacLaughlin and Bill Green, who were on patrol in a Marine Sanctuary boat, fifty yards away. The pair motored over to Walter. While Bill radioed for help, Lauri jumped into the water and helped Walter lift Joanne up on the boat.

Walter stripped off Joanne's diving gear and Lauri checked for breathing and a pulse. Joanne had neither, and Lauri initiated CPR.

Nearby, Joe Marzella was snorkeling with his wife and brother, Mark, when he noticed a commotion on the Benjamins' boat. He climbed into his boat and looked through his binoculars. It appeared the people on the boat were in trouble.

"Do you need any help?" Joe shouted through cupped hands.

"Yes," Bill yelled back. "We need help with CPR."

Joe swam to Mark and told him that help was needed on a nearby boat. The brothers swam to the Benjamins' boat and climbed aboard. From their experience over the years as firefighters/EMTs, they realized the seriousness of the situation and knew that Joanne's chances of survival were slim.

"All I could think," recalls Joe, "was that this poor man had lost his wife right there in front of us, right before our eyes."

Joe swam to the marine sanctuary boat for a tank of oxygen, which they used to ventilate Joanne. Then he suggested they get to shore as quickly as possible. The three boats headed in as Lauri and Joe continued to administer CPR. Joanne had been without a pulse or breath for nearly ten minutes.

"I thought that I may have lost her," recalls Walter. "I didn't think about it as much getting her into the boat as I did on that terrifying ride to shore."

Moments from the end of their twenty-minute ride, the oxygen being pumped through Joanne's system started having a positive effect, and she regained a pulse. The boats docked at the marina where rescue units awaited their arrival.

Joanne was transported to a nearby hospital and treated for cardiac arrest. She was released ten days later. A surgically implanted pacemaker now allows her to lead a more healthy life.

Since Joanne's rescue, the Benjamins have become friends with the Marzella brothers and remain forever grateful to them for helping to save Joanne's life. The memory of that day still haunts Walter.

"I was so scared," he says. "I still get all choked up when I remember what happened."

"I am so lucky to be alive today," says Joanne. "Not a day goes by that I don't think about how lucky I am that all these people were around at the right time."

Teen Bullet

On the afternoon of September 24, 1991, seventeen-year-old Stephen Boland was playing guitar on the front porch of his house in Fairfax, South Carolina, when a friend drove up and stopped to talk. Stephen noticed a gun sitting under the front seat of his friend's car and asked to see it. The friend warned Stephen that the gun, which belonged to his mother, was loaded. Curious and fascinated, Stephen picked it up and aimed it at the ground, then pulled the trigger. Nothing happened. Stephen pointed the barrel at his brother's car and pulled the trigger a second time. Again, nothing happened. Ignoring his friend's warnings, Stephen, confident that the gun was not loaded, aimed it toward his own neck. He fired, and the gun went off.

The bullet struck Stephen in the neck and ruptured his carotid artery. The friend ran into the house and told Stephen's nineteen-year-old brother, Michael, to come out quickly. As Michael rounded the corner of the front porch, he saw Stephen lying on the sidewalk, blood pumping from his neck. He immediately directed the friend to apply direct pressure to Stephen's gunshot wound with his hand and keep it there.

"Mama, you've got to go to Stephen! I've stopped the bleeding!" yelled Michael as he ran inside and dialed their local emergency number.

From the urgency in Michael's voice, his mother, Billie Jo, knew Stephen was in trouble. She ran outside, where Stephen lay unconscious, struggling to breath. As a registered nurse, she knew immediately to lift his legs to keep the blood away from his extremities and from pooling around his vital organs.

"I was paralyzed with fear," she recalls. "I saw all that blood and said to myself, this is my child and he's going to die."

By the time Allendale County rescuers arrived, Stephen was without a pulse and was not breathing on his own. After being stabilized by paramedics, Stephen was loaded into a helicopter destined for the trauma center at Medical University of South Carolina at Charleston. His parents watched the lift-off, painfully aware this might be the last time they would see their son alive.

In the trauma center, Stephen underwent surgery to repair his carotid artery. Doctors were unable to remove the bullet, which had lodged in the spinal cord, and told the Bolands that Stephen would suffer extensive paralysis.

Stephen was hospitalized for three months and today continues to undergo intensive physical therapy. His determination to regain the use of his limbs has paid off, and he now has substantial use of his left leg and minimal use of his right arm.

"I remember I couldn't move anything," recalls Stephen, "and I was so scared it was going to be like that for my whole life. I know it will be a lot of hard work to get it back, and that's what I plan on doing."

Doctors credit Michael Boland for saving his brother's life. They say that had Stephen lost one or two more pints of blood, the results would have been far worse.

"Had it not been for Michael learning how to apply pressure to the wound, which he had seen on *Rescue 911*, Stephen could have bled to death," says Billie Jo. "We'll always be grateful for that."

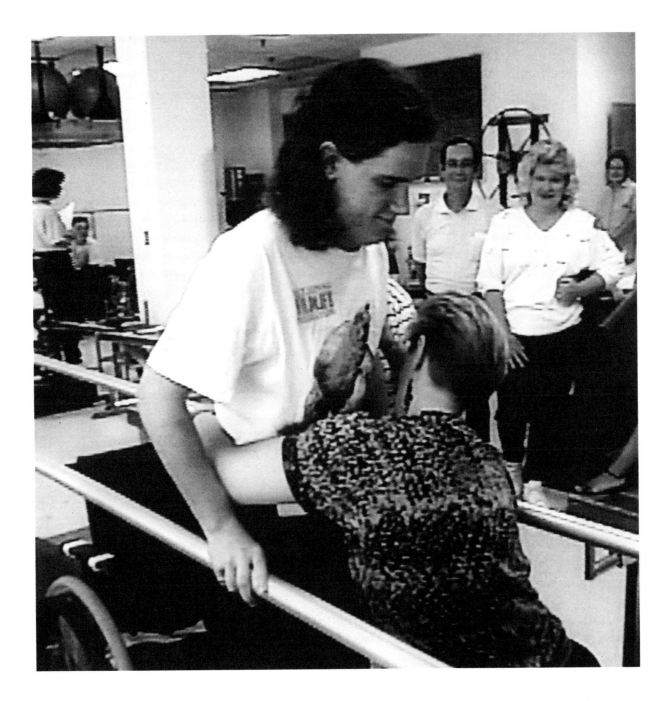

Chatty Robber

Margaret Fisher had been caring for her patient for only a few weeks. A nurse's aide, Margaret would arrive at Richard Holt's residence at night and stay until early morning to care for the fifty-nine-year-old man who was wheelchair-bound as a result of a stroke and leg amputation. Around one-thirty in the morning on December 19, 1989, Margaret was awakened by a loud banging on the door. Frightened, she called 9-1-1.

"My name is Margaret Fisher," she said to Salt Lake City Police Dispatcher Trainee Laurie Anderson. "There's somebody banging on the door and he's trying to get in."

"Okay," responded Anderson, "we'll keep you on the line until the officers—"

"Oh! He's in the house—!"

The line went dead. Anderson had been in training only six weeks and turned for help to her instructor, dispatcher C. J. Prisbrey, who had been monitoring the call. Prisbrey knew it was time to step in and let Anderson listen. Because Salt Lake City had Enhanced 9-1-1, Prisbrey was able to see Margaret's phone number displayed on her computer terminal and called her back.

"Hello?" answered a male voice.

"This is the police department," said Prisbrey. "Who is this?"

"This is Mrs. Holt," the man replied.

"What is going on there?" demanded Prisbrey.

"I think she had a prank call."

"I don't think so," said Prisbrey sternly.

Whoever the man was, Prisbrey wanted to stall him until police officers arrived.

"Your mind is going a million miles a second," says Prisbrey, "trying to think of something to keep him on the phone. If he's wasting his energy talking to me, he's not beating up on someone."

The man told Prisbrey that his mother had called 9-1-1 when he came home because she was confused and didn't realize who he was. He was so stubborn, and pressed the issue for so long, that Prisbrey paused to consider his story for a moment.

"We have little old ladies," she says, "who get very confused, who don't know who their family is. The guy was so insistent, that it was a slight possibility."

Officer Roger Williams responded to what he thought was a family fight and waited outside the Holt residence for backup. On the phone, the intruder continued his story that he was Margaret's son. Suddenly, Prisbrey and Anderson heard a lot of commotion in the background. Then it sounded as if a shot had been fired.

"What is going on!" demanded Prisbrey.

The line went dead. Prisbrey called back again and waited for someone to answer. Meanwhile, police officer Michael Jensen joined Williams at the house. At closer range, Williams could see the suspect inside with a gun and quickly realized a robbery was in progress. The officers positioned themselves outside and drew their guns. When the suspect finally exited the house, they apprehended him.

Margaret answered the ringing telephone and explained to Prisbrey that the suspect, who had stolen eighty-six dollars, was now in custody—thanks to Prisbrey having distracted him long enough for officers to arrive before he fled the scene. Prisbrey learned that the shot had been fired by Richard, who had wheeled himself into the room with his gun and shot at the robber. The suspect wrestled the gun away from Richard and put it to Richard's head. Fortunately, he didn't shoot.

"The suspect was looking for an easy hit for some drug money," says Williams. "If the dispatcher hadn't distracted him with the telephone, the situation could have really deteriorated and the lives of the older gentlemen and woman may have been in danger."

"I've never talked to a burglar before," says Prisbrey. "They don't usually answer the phone. I'm just lucky he decided to pick up the ringing phone and spend some time chatting with me."

Prisbrey praises Margaret for doing a great job despite her fear. She stayed calm, remained on the line, and answered Prisbrey's questions. Margaret says she wanted to run out of the house, but she didn't out of concern for Richard. The next night, she returned to work to care for him.

"She's a lot of comfort to me," says Richard. "She's become a good friend of mine. I look forward to her coming in at night, I really do."

9-1-1 Kids in Smoke

On the morning of August 29, 1991, twelve-year-old Janel McInnes, of Huntington Beach, California, was babysitting her two-year-old brother, Tyson. Their mother, Denise deVines, felt secure leaving her daughter in charge while she attended a business meeting. The children were upstairs, and Tyson was playing with a toy at the top of the stairway. He threw the toy downstairs and Janel told him to go down and pick it up.

Tyson went downstairs for his toy, then wandered into the kitchen where toy trucks were sitting on top of the stove. Tyson stood on a stool and reached for a truck. As he stepped down, his body inadvertently twisted a stove knob, causing a burner to ignite. Tyson went back upstairs, unaware that his other toy trucks were catching fire.

Janel smelled something burning. She looked over the upstairs railing and saw black smoke billowing up from the living room. Janel tried to run downstairs, but the heat drove her back. She pulled Tyson into her bedroom, closed the door, and placed a towel under the doorsill. She knocked out her window screen, but couldn't escape through the window. Then she dialed 9-1-1.

Huntington Beach Fire Dispatcher Mary Ann Marcello answered the emergency call.

"There's a fire in the house!" said the girl on the line. "We can't get out of the bedroom! There's too much smoke! I can't even breathe!"

Janel told Marcello she and her brother were trapped in a second-story bedroom. While Marcello stayed on the line with Janel, she dispatched fire and rescue units to the scene. Janel's voice grew more panicky. Again, she told Marcello she couldn't breathe.

"I can hear it!" she cried to the dispatcher. "Help us! I can hear the fire!"

"Don't open the door," ordered Marcello. "Get down on the floor and stay down."

Marcello was frightened. She'd been a dispatcher for many years and knew that people could perish from smoke inhalation in a matter of minutes.

"Help!" Janel coughed into the phone. "Please help us!"

Marcello listened as Janel coughed, overcome by smoke. She turned to her dispatch partner.

"The firemen have to get into that building," she told her. "But it might not be soon enough."

Paramedic Dave McBride arrived with fire units. Firefighters decided on a simultaneous rescue attempt through the upstairs window and through the front door. When McBride entered from the front, fire was rolling from the kitchen into the living room and visibility was next to nothing. McBride reached the top of the stairs and heard Janel's screams. He and the other firefighters reached the children and carried them out of the burning building. They were frightened but unharmed.

When a report came over Marcello's radio that the victims had been rescued, she was very happy and relieved.

Janel is grateful to her rescuers. She recalls their arrival at her bedroom door.

"I remember the vision of a fireman coming in the door and thinking, this is my hero."

"Janel did everything right," praises McBride. "They could have died, but she remained calm."

Janel says she learned what to do through fire safety lessons in school, and she learned to stay calm through junior lifeguard training.

Ironically, one week before the fire, Janel's mother, Denise, thought of installing new batteries in the smoke detector.

"I feel guilty," says Denise, "when I think how this could have been avoided by changing a two-dollar battery."

The children's father says everyone should know how important smoke detectors are.

"That morning, we needed it."

Heart Attack Runaway Car

On February 14, 1989, sixty-six-year-old Leonard Moran of Jackson, Mississippi, was driving in rush-hour traffic to meet friends for breakfast. Motorist Lynn Porter was driving behind Leonard when she noticed him lean forward in his seat. Lynn figured he was reaching for a car phone, but when Leonard didn't come back up, something told her he might have suffered a heart attack. Lynn stopped and shifted into park, hopped out of her car, and started running down the middle of the street to catch Leonard's car, which was moving toward a busy intersection.

Another motorist, Mark Gallagher, was driving his truck in the lane next to Leonard. Mark also noticed Leonard slumped over the seat. Unaware of Lynn Porter's efforts, he thought he'd better stop Leonard's car before it hit the intersection. Mark pulled in front of the car and sped thirty yards ahead, then stopped, intending to let Leonard's car rear-end his truck rather than roll into the intersection.

Meanwhile, Lynn had caught up with Leonard's car. As she ran, she grabbed hold of the door handle and flung open the driver's side door. Lynn clutched the steering wheel and tried unsuccessfully to set the emergency brake. She struggled to hang on, worried that she would slip under the car. She mustered all her strength and grabbed the steering wheel with both hands. She pulled herself into the car, threw herself on top of Leonard, and slammed the gearshift into park. She jumped out of the car screaming for help and encountered Mark, who had just caught up with her.

Mark told Lynn to run and call 9-1-1 while he tended to Leonard. He pulled Leonard out of the car and laid him in the street, then checked for a pulse and breathing and found neither. Mark had taken a CPR course fifteen years earlier and called upon his memory as he began administering compressions and rescue breathing.

Lynn ran from one house to the next looking for a phone, upset as she pounded on doors, and got no response. Lynn ran into a nearby paint store and yelled for help. Off-duty firefighter, salesman Howard Taylor, raced out to help Leonard while Lynn called 9-1-1.

Firefighters were immediately dispatched to the scene, but the nearest ambulance with advanced life support was five miles away. Howard took over administering compressions, while Mark continued performing rescue breathing. Howard was relieved when EMTs arrived and took over for him.

When paramedics arrived, Leonard was clinically dead, still without a pulse or breath. Rescuers administered a couple rounds of drugs and shocked Leonard's heart, but they had no effect.

"I was very distraught after I realized he didn't have any vitals even after they hit him with those paddles," recalls Mark. "The whole time I was thinking, maybe I could have done better."

Leonard was loaded into the ambulance and transported to the hospital. Howard, concerned about Leonard's condition, called for a status report.

"Man, you did it," he was told.

"Wow. I did it?" asked Howard.

"Yeah, you did it. You did it."

Seventeen minutes after Leonard suffered his heart attack, he finally regained a heartbeat. He was stabilized and breathing on his own. Two weeks later, he was released from the hospital.

Leonard and his wife, Mary Louise, have not forgotten the people who helped save Leonard's life that day. They threw his rescuers a thank-you dinner and gave each of them a heart-shaped plaque.

"Had it not been for them, I wouldn't be here," says Leonard. "They're in my thoughts almost every day. They're very special people to me."

Lynn says that the single most important thing she would emphasize about the incident is the importance of knowing CPR.

"You can stand there and look at the person, or you can give him a second chance."

Says Mary Louise, "What a wonderful gift to get Leonard back on Valentine's Day."

Rocky Raccoon

On November 17, 1990, a Shreveport, Louisiana 9-1-1 dispatcher answered a strange call for help.

"I was wondering," asked the woman on the line, "do y'all help raccoons?"

"Do we help raccoons?" replied the incredulous dispatcher.

"Yeah," said the caller, Rhonda White, "we have this raccoon stuck in the limb of a tree."

Moments earlier, Rhonda's husband, Richard, had been getting into his car parked in the driveway when he heard a loud rustling in the tree above. Richard looked up and saw what he thought was a squirrel, but upon closer examination, he realized it was a raccoon. The animal's head was stuck inside a knothole on one of the tree limbs, and its four legs dangled in the air about forty feet above the ground. Richard ran inside and got Rhonda. The Whites felt sorry for the trapped animal, who was obviously suffering as it fought to free itself, so Rhonda went inside and dialed 9-1-1.

The dispatcher told Rhonda she would have to call Animal Control. Rhonda did, but Animal Control told her they didn't have the equipment to get up that high in a tree. Rhonda called back 9-1-1 and again was told that the fire department could not help.

Determined to help the raccoon, Rhonda called her father, Ted Griffin, a man in his seventies, and asked him to come over with a tall ladder, rope, chain saw, and whatever else it would take to get a raccoon down from a tree. Ted loaded his car with equipment and headed over to his daughter's house.

"We had an emergency going there," recalls Ted, "and I had to get going. When Rhonda calls, Daddy has to go."

Meanwhile, the Whites' neighbor, Reggie Hargrove, was leaving his house when Richard called him over to look at the raccoon.

"It was an absolute pitiful sight," recalls Reggie. "I knew if we didn't get him out he was going to die."

Reggie made a third call to 9-1-1, thinking he might be more persuasive than Rhonda had been. Reggie told the dispatcher that his neighbors were planning to climb a tall tree and that somebody was going to get hurt. That motivated the dispatcher to call Shreveport District Fire Chief Charles Scarborough and ask if he could help.

"They told me there was a coon hanging in a tree," recalls Chief Scarborough. "I said to myself, there's a coon hanging in a tree? That's where a coon lives—in a tree."

Scarborough headed to the scene to check out the situation, as Ted Griffin arrived with his equipment. To Reggie's and Rhonda's dismay, Richard and Ted didn't wait for the fire department—they had climbed the ladder and stood in the tree trying to figure out how to help the poor raccoon, who would alternately growl and snarl then collapse with fatigue.

"It reminded me of Laurel and Hardy trying to rescue a raccoon," recalls Reggie, who thought the men had lost all sense of priority. "I just kept telling Richard, 'get out of that tree, Rambo. You're going to hurt yourself.' And he kept telling me to be quiet."

Chief Scarborough arrived and realized it would take a hook and ladder truck to safely rescue the raccoon. When the truck arrived, Richard and Ted came down, leaving the job to the experts. Scarborough and his men climbed up and decided the safest thing for the raccoon would be to saw off the tree limb, carry it down, then free the creature's head. Once on the ground, firefighters split the log as though it were a piece of firewood, and the raccoon was free. With outstretched arms, one of the firefighters carried the clawing, frightened animal away from the crowd of cheering children who had gathered, then let it go. The raccoon quickly scampered off.

"It went in my yard," says one of the children, "climbed up a tree and went back to his mom and dad."

Richard, Rhonda, Ted, Reggie, and all their neighbors were delighted that the raccoon had been saved.

"We all felt that we had accomplished something good," says Reggie. "The raccoon would live; the kids got a chance to see it. And it was probably the most exciting thing that happened here in Shreveport for the entire month."

Sinking Sisters

In the late afternoon of November 30, 1990, Kelli McWhorter and her three children were about to embark on a trip to Kelli's mother's house for the weekend. Kelli started the engine of her station wagon, which was parked in front of her trailer in rural Conroe, Texas. Twenty-two-month-old Nicole was seated in the front, and three-month-old Whitney was buckled into her infant seat in the back. Brandon, Kelli's six-year-old son, was standing outside by the car. At the last minute, Kelli remembered something: the roast she'd promised to bring for dinner was in the refrigerator. Kelli left the car in the "park" position, the engine running, and ran inside the trailer. Brandon followed, saying he wanted to help.

As Kelli removed the roast from the refrigerator, she noticed through the kitchen window that the station wagon was rolling rapidly downhill, toward a pond twenty feet away. She dropped the roast and ran outside, but she was too late. The car nose-dived into the water and began sinking at a forty-five degree angle. Panicked, Kelli jumped into the pond and swam to the car, where she could see Nicole and Whitney trapped inside. The driver's and passenger's windows were rolled up and the automatic door locks were set.

"Nikki, unlock the doors! Unlock the doors!" Kelli screamed as she pounded on the front window.

But Nicole just stared at her, in shock.

Meanwhile, Brandon had taken it upon himself to call 9-1-1. Dispatcher Willie Roston at Emergency Medical Services answered the call. Roston had to calm Brandon, who was so excited that he was unintelligible.

"The car went down to the lake and the babies are in it!" yelled Brandon. "The car is leaking! The car is leaking!"

When Roston understood what Brandon was saying, he dispatched an ambulance to the general vicinity. He couldn't send paramedics to the McWhorters' trailer because his computer terminal displayed the origin of Brandon's call only as "Highway 150."

Meanwhile, Kelli racked her brain as she treaded water, feeling helpless as she watched Nicole jump into the backseat to escape the water that was filling the front of the car. The thought of life without her girls kept flashing through her mind. Kelli swam to the rear of the station wagon, which was up in the air. The back window was open and the tailgate had never locked well in the past, so maybe she could get inside that way. Kelli managed to unlatch the tailgate, but the car was positioned at such a steep angle that she couldn't hoist herself up inside.

On the opposite side of the pond, neighbors Robin and David Brock were leaving their trailer when they heard Kelli shouting and saw her car sinking. Another neighbor, Randy Welch, was returning home from work when he was also alarmed by Kelli's screams. Brock and Welch ran to the pond and dove into the water. They swam to Kelli and told her to swim back to shore.

Brandon was still on the line with the dispatcher. Through persistent questioning,

Roston finally learned from Brandon the landmarks that would identify the McWhorters' trailer. He informed the paramedics, who were en route in the general direction.

Back at the pond, Brock hoisted himself onto the tailgate. He climbed into the car, grabbed Nicole, and handed her to Welsh, who swam her to safety. Brock went back into the car to look for Whitney. He hadn't seen the baby when he rescued Nicole, and he prayed that she wasn't submerged, especially since the water level was now up to the underside of the dashboard. Brock found Whitney barely above water level, strapped into her infant seat. He pulled her out and had just cleared the station wagon when it disappeared underwater.

Paramedic Linda Bargsley arrived to find that the girls had escaped without injuries. Brandon was ecstatic when he found out his sisters were all right.

"I didn't want my sisters to die, 'cause if I didn't have my sisters I wouldn't have nobody to play with," he says. "The car was leaking and I was thinking that it was sinking and that would be sad."

Roston and his dispatch team also heaved a sigh of relief when they heard the good news.

"Calls like this make you want to run home," Roston says, "so you can hug your children and tell them you love them."

Kelli is grateful to her neighbors, for saving her daughters' lives, and to her son, for knowing what to do.

"I love Brandon with all my heart."

9-1-1 I'm Losing the Baby

Around midnight on May 9, 1992, nine weeks before her baby was due, Cora McElroy of Fontana, California, thought her water had broken. She woke up her thirteen-year-old son, Michael, and asked him to watch his siblings while she went to the hospital for the delivery. Cora, an obstetrics nurse, wasn't too concerned until she turned on the hall light. Then she saw that blood, not water, was gushing out.

"I'm losing the baby!" she yelled to Michael. "Call your grandma!"

As Cora lay down, Michael, who was scared to death by the sight of so much blood, called Cora's mother, Barbara Gullick, an emergency room nurse who lived a few blocks away. Michael handed the phone to his mother. Barbara told Cora to hang up and call 9-1-1.

San Bernardino County Communications Dispatcher Diane Crawford spoke to Cora as a rescue unit was sent to the house.

Cora sounded panic-stricken as she told Diane that she was going to lose her baby. The dispatcher's first impression was that Cora's water had indeed broken, which is sometimes accompanied by blood. She figured that Cora was about to have a normal delivery, but Cora insisted she was hemorrhaging and couldn't feel any fetal movement. When Diane learned that Cora was an obstetrics nurse, she realized this woman had a lot of experience and knew exactly what was happening to her. Diane, concerned that Cora might go into shock as a result of blood loss, instructed Cora to prop pillows underneath her legs and hips in order to keep her body elevated above her head.

"I'm shaking real bad," Cora told her.

That's when Diane became scared for her. She knew shaking was a sign of going into shock. Cora could bleed to death at any moment. Diane tried to calm her by distracting her with light chitchat. She told her to have Michael go outside and flag down the medics, who would be there any minute.

EMTs from the San Bernardino County Agency and paramedics from Mercy Ambulance arrived within moments of each other. Paramedic Mike Richardson saw at once that he had a life and death situation on his hands. He thought Cora might be suffering from *abruptio placenta*, a condition in which the placenta separates from the uterine wall. In 60 percent of these cases, the mother loses the baby. Barbara, Cora's mother, who had arrived in time to watch the ambulance roll away, was keenly aware she could lose both her daughter and grandchild.

Cora was rushed to Kaiser Permanente Hospital, where she worked as a nurse. A medical team, headed by obstetrician Rodney Parker, met her arrival. Dr. Parker confirmed through ultrasound that Cora's placenta had separated from the uterine wall. An emergency Caesarian section had to be performed immediately to save mother and child.

Cora's baby came out blue and limp. He wasn't moving or breathing. Doctors manually pushed air into his lungs through a tube inserted into his body. After five minutes, the baby started to breathe on his own. It appeared he was going to live.

Today, Joshua is a healthy baby, and Cora is doing fine. Michael is grateful that his little brother made such an amazing comeback.

"I love my mother a whole lot, and I'm glad she and the baby are okay."

Cora feels she owes a lot to Diane Crawford.

"She's the reason I have Joshua today. She was so quick in dispatching help. One or two minutes could have made the difference in Joshua being here."

Diane is happy that Cora is so appreciative, but she feels she didn't do anything special.

"I was just doing my job to the best of my ability."

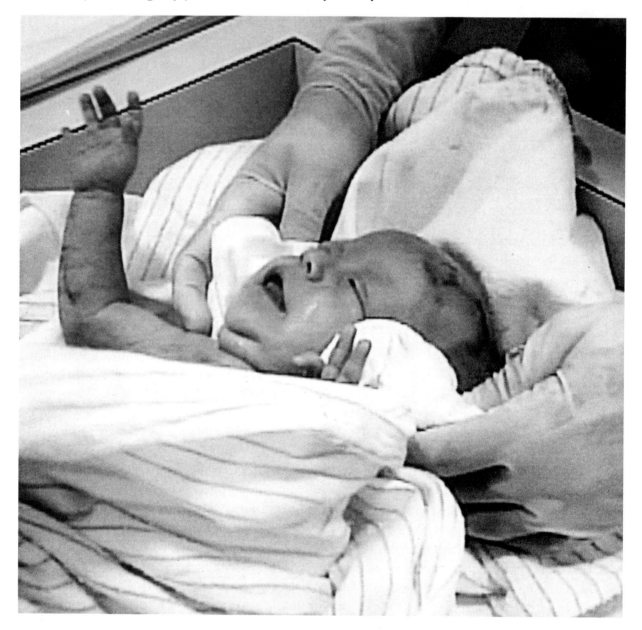

Siblings Save Grandma

Two evenings a week, seventy-three-year-old Mary Cantrell of rural Salcedo, Missouri, has the chance to babysit her great-grandchildren while their father attends a night training class.

On March 10, 1992, David Cooper joined Mary and his children for some of Mary's home-cooked fried chicken. After dinner, eight-year-old Michael and eleven-year-old Mary Beth stayed at their great-grandmother's house while David left to attend his class.

Mary cleaned up the kitchen as Mary Beth and Michael watched television in the living room. Before she loaded her plate into the dishwasher, Mary put one last piece of chicken into her mouth. She swallowed, but the chicken lodged in her throat.

Mary couldn't catch her breath. Frightened, she pounded on the kitchen counter to catch the attention of her great-grandchildren. Mary Beth heard the noise and looked over her shoulder. She saw Mary standing by the counter, choking and gasping for air. Mary Beth ran into the kitchen where Mary was turning blue and clutching her throat. She told her brother to call 9-1-1, but worried that an ambulance might not arrive in time to save her great-grandmother.

While Michael dialed 9-1-1, Mary Beth attempted the Heimlich Maneuver. She had not been trained in the procedure, but she'd seen an episode about it on *Rescue 911* and figured it was what her great–grandmother needed. Mary Beth put her arms around Mary and administered three thrusts without any success. Meanwhile, Michael got a busy signal and hung up.

The children decided to switch places. Michael rushed to Mary while Mary Beth rushed to the phone and dialed 9-1-1. But Mary Beth hung up before a dispatcher could answer. She'd decided it would be better to get a neighbor, so she dashed out the door, leaving a very frightened Michael alone to try to save his great-grandmother.

Michael attempted the Heimlich Maneuver, although he had not received training in it either.

"I was so scared she was going to die," recalls Michael. "I just wrapped my arms around her and started pulling and I just kept doing it. It just flew right out."

Michael had dislodged the piece of chicken on his fifth thrust. When Mary Beth returned with a neighbor, Mary was sitting on the couch, breathing fine on her own.

"My great-grandmother could be gone right now," says Mary Beth, "but she's still with us. If Michael hadn't done that, she probably wouldn't be."

Mary Cantrell is amazed that her great-grandchildren knew exactly what to do and acted so quickly.

"I'm so proud that words can't tell," she says. "Unless someone has saved your life, you don't know how to express it, but I think they're great. They're definitely heroes."

David Cooper, who is, ironically, a paramedic, was also impressed by his children's actions. He was completely surprised at their ability to perform the Heimlich Maneuver without any formal training and says that it has taught him a big lesson.

"Children pick up a lot more than what you think they're picking up on," says David.

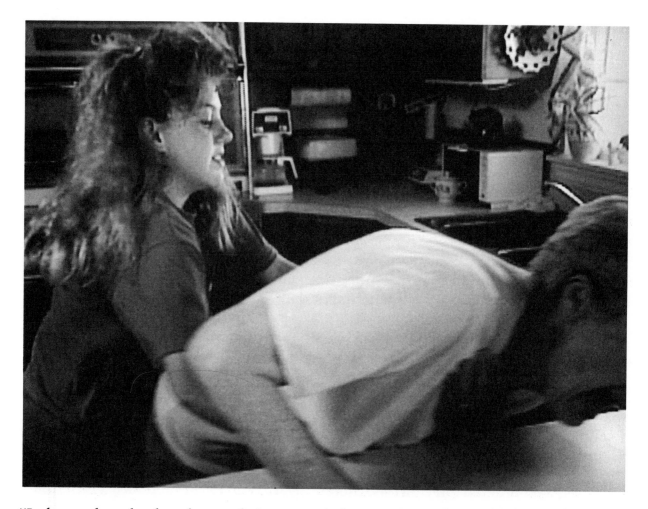

"I always thought that due to their age and the inability to keep their attention that they're difficult to instruct. It's definitely changed my mind on a lot of things."

After fifteen years of experience as a paramedic, David now offers classes on the Heimlich Maneuver to youngsters.

The State of Missouri honored Mary Beth and Michael for their heroism in saving their great-grandmother's life, making Michael the youngest hero on record to be honored by the state.

Potty Peril

On the morning of July 7, 1992, Laura Cole learned that just about anything can happen when children are involved. The Port St. Lucie, Florida, mother of three was holding her baby in her arms while talking on the telephone in the kitchen. Laura's other children, nine-year-old Billy and two-and-a-half-year-old Kristine, were watching television in the living room with their teenage babysitter, Melissa.

Kristine, an active toddler, can sometimes be a little mischievous, according to her family.

"She can be a pain sometimes," says Billy. "She's kind of bossy to me."

Kristine came into the kitchen and told her mother she had to go potty. Laura told Kristine, who was in the middle of being toilet trained, to go into the bathroom, that she would be right in.

Laura was still on the telephone when she heard Kristine crying a few minutes later. She asked Melissa to check on her. In the bathroom, Melissa found Kristine with her left leg in the toilet and the right one hanging out.

"I thought it was strange to find a girl in a toilet bowl," recalls Melissa.

"No way," said Laura when Melissa told her that Kristine was stuck in the toilet and she couldn't get her out. Laura went into the bathroom and pulled on Kristine's foot, but it wouldn't come out. She figured a little lubrication in the form of dishwashing detergent would do the trick, but Kristine's foot still wouldn't budge.

Laura decided to call 9-1-1. She explained to Port St. Lucie County Dispatcher Sharon Murphy that her daughter's foot was stuck in the toilet and she needed help to remove it. St. Lucie–Fort Pierce County Fire Rescue units, headed by Lieutenant Larry Barton, were dispatched to the scene. Kristine's brother, Billy, met the rescuers at the door and led them into the bathroom, where they were greeted by a terrified Kristine, who was screaming vociferously.

The firefighters tried soap, but nothing seemed to loosen Kristine's foot. Never having responded to a situation like this one before, the men needed the advice of a plumbing expert, so Lieutenant Barton radioed his captain, Roger Davis. Barton and Captain Davis agreed that if Kristine could not be removed from the toilet, then the toilet had to be removed from Kristine.

Billy had been concerned about his sister.

"I was worried," Billy says, "because Kristine's only two, and this is the first big adventure she ever had."

At that point, Kristine's father, John Cole, happened to stop by during a work break and found firefighters siphoning the water from the toilet and unbolting it from the floor. Kristine, who had been crying and screaming, calmed down as soon as she saw her father.

While John held Kristine upright, four firefighters carried the toilet out to the driveway. Captain Davis then arrived on the scene and conferred with his men. It was decided they would have to break the toilet. Kristine's foot was wrapped in a towel, and one of the men carefully began to tap the toilet with a tool. The idea was to make cracks in the porcelain so it would fall away piece by piece. Finally, a big piece of porcelain crashed to the ground and Kristine's foot was freed. She leaped into her father's arms.

Kristine escaped from her trauma with only a minor cut. She did not experience any other setbacks in completing her toilet training.

"There's one big lesson to be learned from this," chuckles Captain Davis. "Never stand up in the toilet. And never pull down the handle when you do."

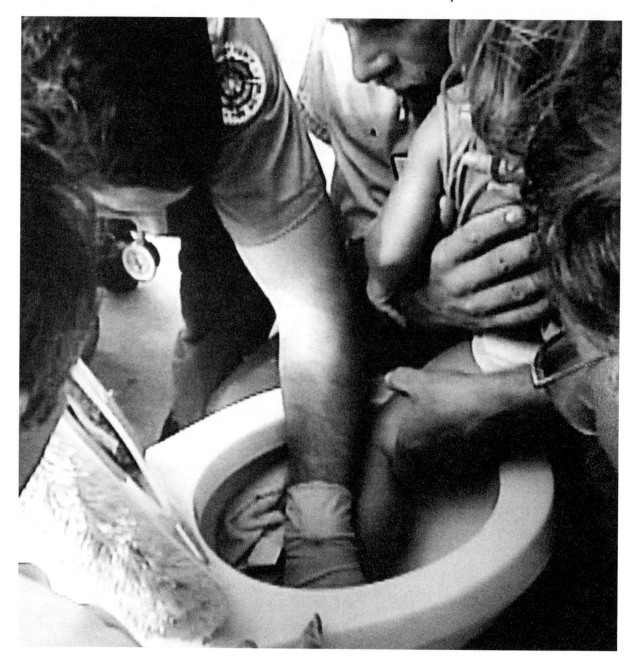

No Helmet Horror

On the afternoon of October 9, 1989, eight-year-old Vicki Swift of Cheltenham, Pennsylvania, and her eleven-year-old brother, Stephen Jr., were at play, chasing each other in and out of the house. When Stephen Jr. took off running down the sidewalk, Vicki jumped on her bicycle and pedaled in pursuit.

Stephen Jr. dashed across the street. Vicki, anxious to catch up, didn't look before she rode off the sidewalk and into the street—into the path of an oncoming van. Stephen Jr. heard the crash. He whipped around and saw Vicki being hurled into the air by the van that had just struck her. Vicki's head hit the curb as she landed on the pavement, and her bike skidded under a parked car and ruptured the gas tank. Terrified, Stephen Jr. ran home screaming.

Off-duty firefighter John Hopper heard Stephen Jr.'s cries for help and rushed out of his house to find Vicki lying unconscious in a pool of gasoline, which was gushing from the broken gas tank. Hopper and another neighbor immobilized Vicki's neck and moved her to safety. Stephen Sr. and his wife, Kathy, alerted by their son's screams, ran to the scene, but a neighbor, afraid for Kathy to see her daughter in such a state, intercepted her before she could reach Vicki.

Fire and rescue units from the Cheltenham Fire Department rushed Vicki to Abington Memorial Hospital, where she was put on a ventilator for her breathing difficulties and was given a CAT scan to determine the extent of her injuries. Doctors told the Swifts that Vicki had a skull fracture, but that she was doing well.

Boy, we lucked out, thought Kathy.

But the Swifts' elation was short-lived. In less than an hour, Vicki slipped into a coma, and the CAT scan revealed that she had suffered a critical head injury.

"We were jubilant about the good news," recalls Stephen Sr., "then all of the sudden they drop a bomb on you."

Vicki was transferred to Children's Hospital of Philadelphia, where Trauma Director Dr. John Templeton learned that Vicki had not been wearing a helmet when the accident occurred. This told him that the entire force of the blow transmitted directly against Vicki's skull and therefore against her brain. Of greatest concern to neurosurgeon Dr. Ann Christine Duhaime was to control the pressure inside Vicki's head caused by the swelling of her brain. Only time would tell if Vicki would survive, and if she did, whether she would ever lead a normal, healthy life.

"I felt like everything just disappeared," recalls Kathy. "And I didn't know, is she going to be normal? Is she going to be all right? It just takes the breath out of you."

After two and a half weeks, Vicki began to emerge from her coma and show signs of improvement.

Says Dr. Duhaime, "Recovery from coma is not like it is in the movies, where suddenly the child sits up and recognizes everybody and asks for a pizza. Recovery is a long, arduous process."

Vicki underwent one year of rehabilitation therapy for speech and motor skills that

had been affected by her brain injury. Remarkably, today she shows no side effects from the accident.

"I was happy that Vicki came out all better," says Stephen Jr. "Well, she's still not the perfect sister. If I had to rate her from one to ten, I'd probably say she's about a nine. There's still things she has to work on to be a ten."

In 1992, Vicki was selected to be the poster child for the American Trauma Society which, along with the National SAFE KIDS Campaign, promotes the importance of bi-cycle safety. Kathy Swift has seen a tremendous change in her small community, where just a few years ago, it was rare to see a child wearing a helmet while bike riding. Today it's commonplace, thanks to bike-safety education.

"The only really good treatment for a brain injury," says Dr. Duhaime, "is to prevent it from happening in the first place."

Prevention means wearing a bicycle helmet that meets safety standards whenever rid-ing a bike.

"Kids balk at helmets, there's no question about that," says Kathy. "But as a parent, that's your responsibility to teach your children and protect your children. No helmet, no bike."

9-1-1 Gunshop

On the outskirts of Fresno, California, it was closing time at Bill's Bait and Tackle shop in the early evening of January 17, 1989. Owner Bill Heasley was removing handguns from the glass display cases for overnight storage in the safe when an armed, masked gunman entered the small shop and began shooting wildly at Bill. Bill was struck by several bullets and collapsed on the floor. A second masked gunman entered the shop, then both men began firing at the display cases, shattering them into thousands of fragments. The men grabbed over a dozen guns and fled.

Bill dragged himself to the phone and dialed 9-1-1.

"I've been shot," he said. "This is Bill Heasley."

Bill told Fresno County Dispatcher Diane Vargas that he'd been shot five or six times in the stomach and legs by a robber. He thought his kidneys were bleeding.

"Take it real easy, okay?" said Vargas.

"I'm not going to make it," Bill told her.

"Yes, you are." But in her heart, Vargas didn't believe her own words.

"Mr. Heasley was not going to make it," she recalls. "I was talking to a dead man."

Police Sergeant Jeff Johnson and paramedic Lori Stephens and her partner, EMT Stephen Braun, responded to the scene as dispatcher Dan Lynch took over Bill's call. Lynch could hear Bill's breathing growing deeper and faster—a sign that Bill was going into shock.

"I'm starting to feel cold." said Bill. "Listen, I don't know if I'm going to make it."

"Yes, you will," Lynch replied.

"Listen," Bill said, gasping for air. "Tell my wife that I love her. Tell my boys that I love them, too . . . and to be good."

The words hit Lynch hard because he knew Bill was on his way out.

"It's like, I'm the last voice between their father and the family," recalls the dispatcher.

Stephens and Braun were first on the scene, but they could only sit and wait in the ambulance, as they had been trained to, for Sergeant Johnson to arrive and secure the shop. Johnson arrived moments later, and he was shocked at the sight of the tiny store, which looked like a war zone. Although Bill was still alive, Johnson knew he was dying. In fact, when he called Stephens and Braun inside, the first thing Johnson did was radio the homicide team.

The medics immediately loaded Bill into the ambulance and rushed him to Valley Medical Center. En route, Bill gave Stephens his phone number and repeated the request he'd made to Lynch—to call his wife and tell her that he loved her and the boys.

Bill was admitted to the hospital in dire straits, having lost more than one-quarter of his blood supply. As he was rushed into surgery, Stephens kept her promise and called Bill's wife, Jill.

"That phone call was awful" recalls Stephens. "I was so scared. I had no idea what to say."

Stephens told Jill that Bill was at the hospital.

"Did he have a car accident?"

"No," replied Stephens.

"Then, was he shot?"

"Bill wants me to tell you he loves you. And the kids, too."

Jill says she's relived many times in her mind what it must have been like for Bill to lay wounded in his store, knowing he was dying.

"When Bill said in the 9-1-1 call that he loved me," says Jill, "it meant more to me than anything he's ever given to me."

Incredibly, Bill pulled through the surgery to repair his gunshot wounds, the most critical of which was from a bullet that entered his stomach and penetrated his kidney and liver.

Today, Bill has completely recovered, although five bullets remain in his body. Both he and his family truly appreciate his remarkable survival.

"Bill's my world," says Jill. "I love him very much. He's made my life very happy."

"I think what kept him alive," says one of Bill's young sons, "was his will to live and his will to see us grow up. I really love my dad a whole lot."

The suspects involved in the shooting were arrested within one week and were subsequently convicted and sentenced to prison.

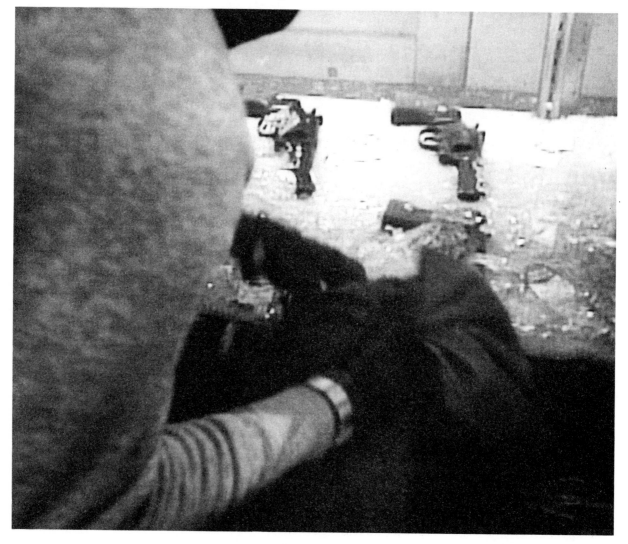

Water Monitor Mayhem

On July 5, 1992, Chris and Maria Elena Hernandez, of Miami, Florida, were backing out of their driveway in Maria Elena's car when the steering column locked. They were on their way to a friend's birthday party, so Chris decided to drive his car and check out the problem later.

When they returned home that evening, Maria Elena, who was eight months pregnant, went inside while Chris popped open the hood of his car and poked around. Chris peered deep into the engine and was stunned. There was a dead snake inside the engine.

Chris went inside and told Maria Elena. He grabbed a broom and they both went outside. Frightened, Maria Elena stood away from the car. Chris looked more closely at the snake and saw a claw grasping the oil pan.

"That's no snake!" Chris shouted. "It's an alligator!"

Chris poked the reptile with the broomstick and the creature reacted. Chris and Maria Elena jumped back. Whatever it was, it was alive.

Chris called the police department on a non-emergency number. The operator, thinking he was a crank caller, gave him phone numbers to various wildlife agencies. When Chris couldn't reach anybody, he beeped his best friend, Tim Lanham, a 9-1-1 supervisor for the Miami Fire Department. Tim was amused by Chris's dilemma—he could just imagine Chris freaking out over a two-foot lizard—but he agreed to help.

When Tim arrived he was shocked to discover the scale of the problem. The creature was completely wedged inside the engine. Tim called the Florida Fresh Water Fish and Game Commission, which dispatched Florida State Police Officer John Esslinger to the scene.

Esslinger thought the creature was some type of exotic lizard and figured he could be dangerous. He contacted Todd Hardwick, who runs Pesky Critters Relocation, a company specializing in nuisance wildlife control. From Esslinger's description over the phone, Todd said the creature might be a water monitor, a large carnivorous reptile in the lizard family that's capable of inflicting a nasty bite. He told Esslinger to be very careful, and he would be right over.

When Todd and his associate Jill Voight took a look at the situation, they excitedly confirmed that the reptile was indeed a water monitor. Jill slipped under the car and tugged on it, but the creature was really stuck.

"All I wanted was to save this lizard and bring it home with me," recalls Jill. "That's what I wanted—the lizard."

As neighbors came over with beach chairs to watch the show, Todd poured liquid detergent over the monitor in the hope he'd slide out.

"But basically after twenty minutes," recalls Tim, "we had a slimy, trapped lizard."

Next, Todd poured a bowl of ice water over the monitor's head, hoping it would react to the cold and back out. Instead, the lizard inflated, further trapping himself within the engine.

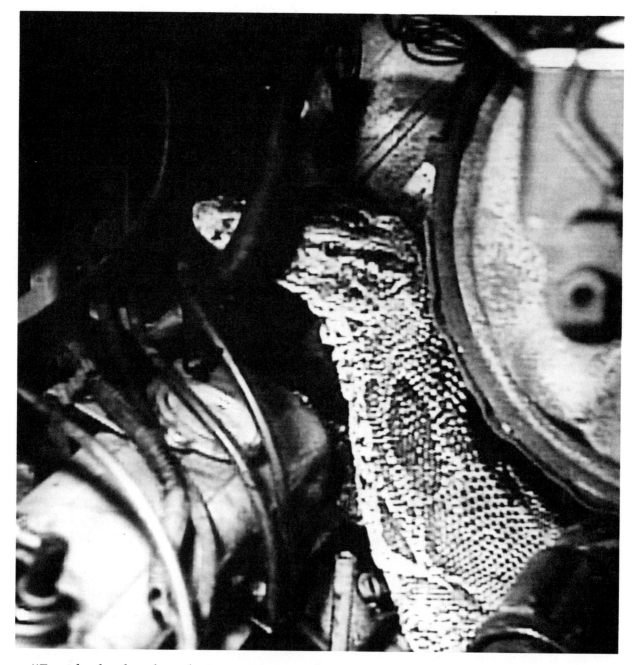

"Everybody else thought it was time to throw in the towel," recalls Todd, "but we don't do that. We get them out alive."

Todd consulted with an exotic animal veterinarian who recommended tranquilizing the monitor. Todd injected it and within one hour, the creature relaxed. Tim slid under the car and unbolted the steering column that blocked the monitor's way. Immediately, the monitor began to drop. Frightened, Tim scrambled from under the car and Jill slipped under to catch her prize lizard. She emerged excitedly hugging the monitor.

"It looked like her long lost pet puppy or something," recalls Todd. "It was like, at last, I've got my hands on you."

They didn't have a cage large enough for the six-foot, twenty-five-pound lizard, so Jill held it on her lap while Todd drove to their home.

The monitor rapidly recovered from its lacerated tongue and minor chest burns and was adopted by a local film company. They named it Gordon Gecko and built it a lush, tropical outdoor cage, complete with waterfall and pond.

"He lives better than any lizard in town lives," says Todd.

"On a food budget that surpasses mine," adds Jill.

"I like Gordon and I'm glad he's fine," says Maria Elena. "And it was fun and adventuresome. At the same time, I don't ever want it to happen again. Not in my car, anyway."

Niagara Falls Rescue

Each year, millions of visitors flock to Niagara Falls, drawn by its spectacular scenery. But over the years, the falls have claimed many lives. About a mile upstream, a set of control gates stretches across the Niagara and regulates the flow of water to the falls. Any boat or person who goes beyond this dam, or "point of no return," will be carried by powerful currents over the falls.

On August 24, 1988, Janelle Spain and Karen Himmel were vacationing at a camp-ground upriver from Niagara Falls. About one to two miles away from the falls, Janelle and Karen cruised the waters in Janelle's brand-new motorboat. They watched the sun-set, then turned back in the dark, thinking they were heading into the canal that led to their campsite. Unfortunately, they had taken a wrong turn.

Sheila and Tom Hodges were driving to a movie on a road that parallels the Niagara River when they spotted a single-engine boat motoring on the prohibited waters toward the falls. Realizing the boaters were in trouble, they stopped their car and saw the boat plunge over the point of no return. Sheila was amazed to see it land upright.

"How that happened—I was just astonished they were still in the boat."

Tom flagged down a park tour bus in time for its driver to see a red flare go up over the water. The driver radioed the Niagara Parks Police Department. Police Sergeant Stan Sherar and Constable Alex Tacinelli were immediately dispatched to the scene.

Meanwhile, Karen and Janelle put on their life vests and jumped from the boat. The Hodgeses watched the women fight the swift current carrying them to the falls.

"Once you pass the international control gates," says Constable Tacinelli, "you have no chance of getting out of the water before going over the falls."

Sherar and Tacinelli spotted Janelle and Karen less than five hundred yards from the falls. Sherar panned his flashlight back and forth on the women while Tacinelli yelled to them to swim toward the light. Karen swam across the current, but the officers lost sight of Janelle.

Recalls Tacinelli, "I turned to my sergeant and said, one's gone. When I said that, I meant one is going over the falls."

Alex threw a rope to Karen. She grabbed it and Tacinelli pulled her from the water. Karen had a deep gash in her leg from the boat's propeller and was in shock, but she was safe.

"I was shaking and I was scared," recalls Karen. "I yelled, where's Janelle? And they wouldn't answer me. I thought, she went over."

Rescuers from the Niagara Falls Fire Department, including firefighters Bob Bevington and Bob Carpenter, joined in the search for Janelle. They drove farther downstream and spotted her 350 yards from the brink of the falls. Janelle yelled that she was tired and couldn't swim anymore.

"I don't give a f - - - if you're tired!" screamed Carpenter angrily, hoping to shock Janelle into keeping up her struggle. "You're going to start swimming to me! I don't give a damn if you're hurt! Swim to me now!"

Janelle continued to fight the current, but the rope wouldn't reach her. Bevington tied the rope around his waist and jumped into the treacherous waters.

"I'm not a brave guy," Bevington remembers thinking to himself. "I thought, this is crazy, what are you doing? Your mother raised an idiot here."

Bevington swam fifty yards across the current to Janelle, grabbed on to her life jacket, and yelled for Carpenter to pull them to shore. On land, exhausted and tearful—but unharmed—Janelle thanked her hero.

"She threw her arms around me, and she crushed me, practically," says Bevington. "It was the most incredible feeling I have ever experienced in my life."

Janelle and Karen cannot thank their rescuers enough. Janelle credits Bob Carpenter for pushing her when she was ready to give up and praises Bob Bevington for risking his life to save her.

"We need more people like him," she says. "I wouldn't be here today if it weren't for him. I don't care if somebody says that's his job. No, that's not his job. He didn't need to do that."

Venezuela Cave Save

On July 13, 1991, Gustavo Badillo Arenas and his friend Eduardo Wallis of Caracas journeyed into the Venezuelan rain forest to explore an underwater cave known to the locals as "Acarite." Accompanying them were Gustavo's fiancée, Maria Elena Mendoza, and a guide, National Parks Representative Asmel Palencia. Maria Elena was not happy about the expedition because Gustavo and Eduardo, experienced open-water divers, had not been trained in cave diving.

At the cave entrance the men floated a candle as a signpost. For a safety line, they attached one end of a hundred-foot rope to a float and the other end to themselves. As they said good-bye, Maria Elena handed Gustavo a prayer card to take with him. The two young men dove underwater and entered the first passage, thirty feet below the surface.

Within minutes, Eduardo and Gustavo became separated in water so murky that visibility was less than an inch. Eduardo couldn't see Gustavo, but he shouted to him to turn back. Gustavo shouted back okay. Eduardo hit one dead end after another but was finally relieved to see the candle glowing in the entrance. When he surfaced, Gustavo was not with him, nor did he respond when Eduardo shouted to him.

Maria Elena insisted Eduardo go back to find Gustavo. The safety line had become untied from the float and was lost in the cave, so the group made a fifteen-foot makeshift line from belts, socks, and scrap rope.

Eduardo reentered the cave, scared for his own life. The murkiness of the water made it impossible to see anything so he turned back. When he reemerged alone, Maria Elena was horrified.

"My life just stopped right there," she recalls. "I thought, this is the end. He's going to die and I can't do anything from here."

By now, it was too dark to send for help, but the next morning, Eduardo and Asmel left to call Gustavo's friend, Vivian Indriago, the owner of the dive shop where Gustavo worked. Vivian knew she wasn't experienced enough to enter the cave, so she called Steve Gerrard in Florida. Steve, one of the most experienced cave divers in the States, and John Orlowski, a river cave specialist, agreed to make the ten-hour trip. They assumed they were flying down to recover a body. In all the years he'd rescued cave divers, Steve had never pulled anyone out alive.

When Gustavo's good friend Leo Calligaro heard the news, he flew in to attempt a rescue. Leo entered the cave. But he immediately turned back, realizing the danger. Devastated, Maria Elena decided to return to Caracas so she wouldn't have to witness the recovery of Gustavo's body.

Thirty-six hours after Gustavo had been lost, Steve and John arrived and entered the cave.

Who in their right mind would ever dive in here?, thought Steve as he swam through the black waters. John didn't like it any better. He just wanted to get the job done and get out of there.

"Why don't we just sit here for two hours and go out and tell them we couldn't find him," Steve said to John at one point, "because we're not going to find him in this."

Three hundred feet into the cave, John saw a light flashing as he swam inside a chamber. He thought he was swimming toward Steve, but when he surfaced, he saw Gustavo coming toward him. the rescuers were shocked. Gustavo was alive. They escorted him out of the cave, where his elated friends greeted them with joyous applause.

Gustavo's problems had begun when the rope came loose and tangled around his legs. He lost all sense of direction and swam in circles until he reached a chamber in the cave. A mud island in the middle saved him. He sat and waited, hoping someone would find him. Eventually, he saw a light and thought the angels were coming to get him. The angels were Steve and John.

"Oh, my God," thought Gustavo when he heard his rescuers' voices, "Heaven is English and the angels are English."

"Gustavo is extremely lucky to be alive," says Steve. "What he did was foolish because he didn't have the right training or the right equipment."

Gustavo agrees. But he plans to return to Acarite after he becomes certified in cave diving.

Maria Elena and Gustavo married one year later and remain grateful to their friends who helped save his life, especially Steve and John.

"That Steve and John would come down from the States to rescue someone they didn't even know—that's a miracle," says Gustavo. "Thank God my angels came and saved me."

Ottawa Bank Bust

The morning of March 11, 1992, was business as usual for a bank in Ottawa, Ontario until a man carrying a knapsack approached a teller, pulled a sawed-off shotgun out from under his coat, and aimed at the teller, demanding cash.

"We have all kinds of robbery seminars," says teller Jeff Hamilton, "and they always told us, don't be the hero. Let them run the show."

That's exactly what the teller did. He filled the robber's knapsack with bills, while an unsuspecting customer, Kevin Mulligan, stood nearby filling out a deposit slip. The robber and Mulligan exchanged looks, then the gunman aimed at him. Mulligan stared back with a defiant look, then obeyed the gunman's order to stand still.

The gunman strolled out of the bank. Mulligan ran out, jumped into his truck, and called 9-1-1 from his cellular phone.

"We have a bank robbery. I have the suspect," reported Mulligan as he tailed the gunman in his truck.

"You have the suspect?" asked police dispatcher Rose Marie Pineo.

"He's running on foot. I'm chasing him in my vehicle. He's going to fire at me."

"Stay away from him," ordered Pineo, afraid that Mulligan was in over his head and would get hurt. "Don't go close."

Mulligan followed the gunman for several blocks, then reported in frustration that he'd lost him. Suddenly, the gunman jumped from behind a building, charged Mulligan's truck and fired several rounds into his windshield.

"He's shooting at me!" Mulligan shouted into the phone.

"Get away from him!" ordered Pineo. "Drive away!"

"I'm trying!"

Pineo was scared to death for Mulligan, who was determined to continue the chase. A police officer met up with Mulligan at an automotive shop where Mulligan had seen the suspect run inside. Moments later, he was joined by several other officers, including Constable Mike Ryan, who led the pursuit.

The suspect fled out the back of the building and led officers on a foot chase, over walls, across a busy freeway, and through a hole in a fence. Then the gunman stopped. To Constable Ryan, he seemed psychotic or combat-trained because, with absolutely no cover, he fearlessly fired upon the officers. They returned fire with handguns, insufficient fire power in this situation because they were too far away.

The gunman ran inside another building and took two employees hostage. A SWAT team surrounded the building, and Inspector Alain Methot was called in to negotiate the release of the hostages, Bob Ireland and Marilyn Bosch.

"This guy was like a bomb waiting to explode," recalls Inspector Methot, who could not satisfy the agitated gunman's demands without endangering the hostages.

The suspect was taking his hostages upstairs when Ireland pushed Bosch and ordered her to run. Ireland spun around and kicked the gunman in the head, knocking him down the stairs. Then he wrestled the gun away from the suspect, ran outside, and shouted to officers to get their man.

The suspect was apprehended in the back of the building in a stolen car. He subsequently pleaded guilty to four felony charges and was sentenced to twelve years in prison.

Kevin Mulligan and Bob Ireland were honored by the Ottawa Police Department for their heroism.

"If Bob hadn't done what he did," says Marilyn Bosch, "neither of us would be here today. The man said he would kill both of us and himself, too."

"He was totally irrational," says Ireland. "A man without a plan in total desperation is very dangerous. I felt I had no choice, no option, and faced with that, I decided to fight."

Mulligan doesn't regret chasing the suspect, but he realizes he could have been killed trying to catch a crazy man.

"Instead of being gutsy," he says, "it was actually nutsy."

"I wish more people would get involved in helping us out," says Detective Mark Pino, who was involved in the pursuit. "But not to the point where they get shot at or hurt, because we're getting paid to confront these people. That's our job. That's what we're trained for."

Racehorse Rescue

On January 27, 1992, horse trainer Pat Mahoney was transporting racehorse Setting Limits and a colt named ML Pursuit to a farm for the winter. The ground in Independence, Ohio, was covered with light snow as Mahoney traveled down the highway in his pickup truck pulling the two-horse trailer.

As Mahoney drove, the horses shifted their weight enough to throw the trailer off balance. Mike and Susan Flores, who were driving behind Mahoney, saw the truck and trailer start to weave and were afraid an accident was about to occur. A second later, Mahoney lost control, and the truck and trailer spun in a circle across lanes of traffic. As they came to a stop on the highway, the trailer broke loose from the pickup and flipped over on its side.

The Floreses stopped and ran to the trailer where the panicked racehorses were trapped inside. Setting Limits was pinned under the colt, and she was bleeding from the colt's kicking in an effort to free herself. Mahoney didn't know what to do. He was afraid to let the horses loose on the highway, but he worried that if left inside, Setting Limits would be further injured. Suddenly, the colt broke free and galloped down the highway's median, but fortunately he was caught one mile away.

Rescue units from the Independence Police and Fire Departments responded to the scene, including paramedic Frank Kruzewski. Rescuers had the equipment to cut the trailer in half, but they didn't know what they would do with Setting Limits if she were released. Although the filly was obviously hurting herself in her struggle to gain freedom, they decided to wait until veterinarian Dan Wilson arrived. Mahoney was most concerned about the filly's bleeding legs.

"All the kicking in the trailer could have done extensive damage," says Mahoney. "They can kick a board and end a career."

Dr. Wilson arrived twenty minutes later.

"A horse's first instinct is to run," says Dr. Wilson, "and when they can't, they panic. A lot of well-meaning people have been killed by horses in a panic situation."

As firefighters cut through the metal bar that was trapping Setting Limits inside the trailer, she collapsed from exhaustion. Dr. Wilson had to get her out quickly to see if he could save her. She was in shock, had lost a great deal of blood, and had a severed artery above her left eye. Mahoney knew she would never race again, but hoped she'd just pull through.

Dr. Wilson didn't want to risk moving Setting Limits because she was in shock, so he got to work on the scene. Kruzewski was amazed to watch the veterinarian perform surgery under such conditions—kneeling on the snow-covered ground in bitter cold to suture the filly's severed artery. Wilson's assistant inserted an IV in the horse's neck.

"The vet told me, just hold pressure and push it in as fast as you can," recalls Kruzewski. "We're not used to pushing so much fluid into a person. They just don't teach us veterinarian medicine in paramedic school."

After he finished the operation, Dr. Wilson held Setting Limits on the ground. Unexpectedly, she sprang to her feet, a sign that she was feeling healthy and strong.

"She just kind of pitched me around like a peanut," recalls Dr. Wilson. "Once she stood up, she never looked back."

Mahoney was beginning to think there might be a happy ending to the incident. And there was. Setting Limits and the colt shocked everybody by walking straight into a new trailer without a moment's hesitation.

"The paramedics and firemen who were there were absolutely great," says Dr. Wilson. "They're the ones who really deserve the credit for saving this filly and colt."

Setting Limits's owner, Shirley Girten, knew her filly's racing future was uncertain when rehabilitation began. Despite all odds, eight months after the accident, Setting Limits returned to the track and won her race. Today, she has retired from racing to have babies.

"She went out in style," says Shirley. "She went out winning just as happy as the day she came in. She's going to have babies, well-deserved babies, I hope, that shall be the runners and carry the heart and stamina that she did."

Falling Glass

Craig Hoffman was on his usual shift as a glass cutter in Portland, Oregon, on the morning of December 23, 1991, assisting co-worker Rich Adams in the dangerous operation of cutting large sheets of raw glass. As the supply on the cutting table neared its end, Rich sent Craig for a new case of glass. Craig, clothed in protective leather, approached the ten-foot cases, each filled with two tons of untempered glass sheets measuring six by eleven feet. Using wire cutters, Craig cut through the first of four metal safety straps that securely held the sheets in place. But the sheets of glass had been incorrectly balanced, and they toppled forward to crash on top of him. The impact of four thousand pounds of jagged shards hurled Craig backward, causing his head to slam into an opposing case of glass, which now dangled precariously above him.

The deafening shatter of glass brought the factory to a standstill. Rich Adams and other co-workers rushed to Craig, who was conscious and struggling to push the glass off his body. Plant manager Larry Henry noticed the commotion on the floor and ran over. By the time he scrambled on to the mountain of shards to begin digging, Craig had lost consciousness.

A call for help sent rescue units from Tualatin Valley Fire Department and Buck Medical Service to the scene. While they were en route, Craig's co-workers, including acting supervisor and old friend Kevin Smith, joined Henry in digging furiously to extricate Craig from the glass.

"My heart dropped right out when I saw Craig," recalls Kevin, who was worried that Craig would be dead before rescuers could get to him.

Within five minutes, fire rescue units arrived and radioed for a LifeFlight helicopter to be put on standby. Paramedic Gene Ditter initially thought Craig stood little chance of surviving. He was without a pulse and wasn't breathing. Paramedic Tom Duthie administered CPR, although he also felt Craig was not going to make it, considering that less than one in one hundred trauma patients without a pulse can be revived.

Because Craig's jaw was tightly clenched, paramedics were unable to administer a flow of oxygen until LifeFlight arrived. Flight nurse Sue Geleski injected Craig with a drug that causes temporary paralysis, allowing Ditter to open Craig's jaw and insert an oxygen tube. Co-workers stood by in despair as their lifeless friend was loaded into the helicopter.

Meanwhile, Craig's wife, Jean, received news of the accident at home. She rushed to the hospital, knowing the situation was critical when she heard the name LifeFlight.

Craig was admitted to Oregon Health Sciences Hospital and underwent surgery to repair a ruptured lung and numerous fractures to both legs. His prognosis for survival was grim, due to extensive internal injuries.

Remarkably, as the days passed, Craig's condition began to improve. Three weeks later, he was discharged from the hospital without any brain damage, an amazing development considering he had been without oxygen for eight minutes. After six months of rehabilitation, he returned to work—in an office job.

Dr. Donald Trunkey attributes part of Craig's survival to his co-workers' and rescuers' quick actions.

"If any credit is due, it's to the people at the scene. They did a fantastic job," praises Dr. Trunkey.

Jean Hoffman feels very lucky that Craig is alive, and he concurs.

"I think it's a miracle," he says. "I'd like to thank a lot of people for praying for me."

Phoenix Flood

When a series of summer thunderstorms hit Phoenix, Arizona, in 1992, Phoenix Unified 911 Dispatch was prepared for the incoming emergency calls that usually accompany this desert area's flash flooding. On August 22, around eleven o'clock at night, dispatcher Amy Holtz received several calls reporting a motorist trapped in her car, which was lodged against a pile of rocks in a flooded riverbed. The stranded motorist was thirty-five-year-old Ngan Tran, and one caller anxiously urged Holtz to get help to her immediately, since water had already reached the hood level of her car.

Members of the Phoenix Fire Department Technical Rescue Team, including Firefighter Ron Cummings, were dispatched to the scene. Phoenix Police Aviation Officers Sallie Scott and Mike Hein, who were on routine helicopter patrol, also headed for the area. When the rescuers arrived, twenty-five-miles-per-hour rapids had risen to the roof of Tran's car.

"My first thought was, we don't have a rescue here," recalls Cummings, "we have a body to recover." Even if rescuers could get to Tran, Ron thought they'd either die in the process, or she would be dead by the time they got to her.

There was no time for elaborate planning. A second helicopter was called in to attempt the rescue as the first chopper hovered to illuminate the car with its spotlight. Cummings would be lowered from the helicopter's skid to the roof of the car to try to reach the trapped victim. Downstream, safety lines were set up at intervals across the river, in the event that Cummings and Tran were swept away.

Cummings donned a wet suit and life preserver and mounted the helicopter's skid. Pilot Larry Hallas took off and flew toward the car, then hovered as low as he could over the rushing water. Without any steady reference point, it was nearly impossible for Hallas to determine if his helicopter was drifting or on target. But co-pilot Carl Hagler assisted him in positioning the skid directly above the roof of the car, which now had only about eight inches of dry surface. Cummings stepped onto the roof, and the helicopter retreated.

In a prone position atop the car, Cummings leaned over to look inside it. Tran was sitting in the backseat, up to her neck in water. Cummings broke the rear window and motioned for her to stay put, then he summoned the hovering helicopter for a helmet and life preserver. Cummings helped Tran put on the gear, climb out the window, and crawl up on the roof of the car.

In a delicate maneuver, Hallas piloted the helicopter back a third time. He had to position the aircraft so the skids were just within Cummings's reach, but at the same time he had to avoid knocking him and Tran into the water. The problem was compounded by the electromagnetic charge that a helicopter generates when it hovers over water. Each time Cummings grabbed the skid he received an enormous shock. After several failed attempts, Cummings finally boosted Tran and himself onto the skid. With hundreds of pounds of extra weight on one skid, Hallas carefully lifted the tilting, unbalanced helicopter into the air, boarded Cummings and Tran, and flew to rescue vehicles waiting on the riverbank.

Tran was treated for mild hypothermia and was able to return home that night.

Ngan Tran remembers being trapped in her car, knowing she was going to die, and thinking of her ten-year-old daughter. Tran says she's learned a major lesson as a result of her ordeal. *Earlier on that evening, she'd encountered a police barricade set up at a flood point to prevent disastrous incidents such as hers.* Since the water level was only up to her tires, she had driven around the barricade. Tran realizes now she risked not only her life but also the lives of her rescuers.

Tran and her daughter, Lynn, paid a visit to her rescuers to say thanks.

"It was a team effort," says Cummings about his most difficult swift-water rescue to date. "If one part had failed, I don't think she would have lived."

"Thanks is not enough," says Tran. "I will remember Ron Cummings forever in my life. And maybe in the next life and the next life."

I've Been Shot

On the night of August 19, 1991, Pat Douglas of Indianapolis, Indiana, drove her thirty-two-year-old daughter, Portia, home after they had gone out for dinner. While Pat waited in her car for Portia to go into her apartment and get a hat Pat wanted to borrow, she noticed a man pacing nervously behind the window in his ground-floor apartment. Pat thought it was odd but gave it only a moment's thought as Portia returned with the hat and said good night.

As Portia checked her mailbox before going to her apartment upstairs, the same young man, who was new to the building, approached her and asked to borrow some salt. She entered her apartment, leaving the door open. She got the salt and returned to her living room, where the man stood waiting, pointing a gun at her. Portia screamed for help as the man fired, hitting her directly in the chest.

Portia lay on the floor and dialed 9-1-1. At 9:04 pm Marion County Sheriff's Department Dispatcher, Victoria Samuels, answered the call.

"I've just been shot twice by a rapist," Portia whispered. Her breathing grew more labored as she gave the operator all the pertinent information.

Sheriff's Department Officers and rescue units were immediately dispatched to the scene, but Samuels was concerned that Portia might die before help arrived.

"I really almost lost it," recalled Samuels, "when she told me to call her parents in case she died. I thought, I really don't want to have to make that call."

Lawrence Township Fire Department paramedic Jeff Muszar, stationed across the street from the apartment complex, arrived within sixty seconds. Protocol dictated, however, that he not enter the building until law enforcement secured the area. He anxiously waited in his vehicle, frustrated that he might not get to Portia in time, but knowing if he went in, he risked being shot by the assailant.

Meanwhile, Amy Waters answered another emergency call from a man who reported he'd heard gunfire and noticed that his gun was missing. When Waters learned that the caller's address was the same as Portia's, she instantly knew it was the suspect, who must have assumed Portia was dead.

Officers, including Sheriff's Deputy Kelly Hayes, who were en route, were informed that the suspect was in his own apartment. Their job now was to get Portia out of the building without letting the suspect know they were on to him.

When the officers arrived, Deputy Hayes went straight to Portia's apartment to be with her until the paramedics arrived. Then the other officers ordered the suspect out of his apartment and took him into custody without further incident. He was shocked to hear that Portia was still alive.

Finally, paramedic Muszar was allowed to enter the building. In twelve years of service, it was the longest wait he'd endured. Portia was rushed to the hospital with tremendously low blood pressure and internal bleeding. Sergeant Dave Tilton arrived on the scene and got a confession from the suspect.

Portia was hospitalized for ten days, then released.

"What happened to me was terrible," she says, one month later and back to work as an attorney, "but I now know I'm a survivor, and I've gone on with my life. I work hard, I play hard, and I do everything as if today were my last day."

Pat says that her daughter has a lot more confidence today and she's proud that Portia showed that night what she had in her.

"I think more than anything, telling the story is very purging," says Portia. "I feel sorry for rape victims that have to hold it in. If you hold it in, it just eats away at you. If you can tell it and get it out, you can go on with your life."

Portia has nothing but praise for her rescuers, especially dispatcher Samuels.

"I was terrified that night," Portia says. "But nothing was as soothing as hearing that operator's voice on the telephone. It was my line to reality."

DUI Teen Driver

Shortly after midnight on March 1, 1987, Brandon Silveria got behind the wheel of his car and embarked on a drive that would forever change his life. Brandon and three crew teammates from Los Gatos, California, were returning home from a friend's house where they had been partying and where alcohol had flowed freely. Brandon was intoxicated as he drove, and so were his friends, all of whom nodded off to sleep. But they were soon jolted awake when Brandon dozed off, too, and sideswiped a guard rail. One by one, Brandon dropped his friends at home, then headed to his own house.

Brandon fought to stay awake as he traveled the highway at fifty miles per hour. But one mile from his house, he fell asleep again. His car veered into the median and crashed into a tree, causing Brandon to hit his head on the windshield, even though he was wearing a seat belt. The car rebounded onto the highway, where an approaching driver, unable to see Brandon's disabled car now that it was without taillights, crashed into it. Brandon's head hit the windshield again.

A nearby resident heard the crash and called 9-1-1. Rescue units from the Santa Clara County Fire Department responded to the scene. When EMT Joe Viramontez saw the wreck, he figured he might have a fatality on his hands. He found Brandon unconscious and assumed he had a head injury due to the spiderweb cracks fanning the windshield. Brandon's blood pressure readings confirmed massive brain swelling.

Within seven minutes, advanced life-support arrived. Paramedics quickly loaded the critically injured Brandon into the ambulance and transported him to San Jose Medical Center.

Brandon's friend Jay was driving to his girlfriend's house shortly after Brandon dropped him off that night when he happened to pass the accident. He recognized the personalized license plate on the demolished car. Horrified, Jay pulled over and spoke to Deputy Bruce Rak of the Santa Clara County Sheriff's Department. Deputy Rak asked Jay if Brandon had been drinking, then asked him to notify Brandon's parents about the accident.

"It was the hardest call of my life," recalls Jay. "I was feeling very guilty that I hadn't stopped him from driving home."

Brandon's parents, Tony and Shirley, were in bed when they received what Tony describes as "the midnight call you always dread when your kids are away." Tony and Shirley drove to the accident scene. From the looks of things, they wondered if Brandon was still alive.

After Brandon was admitted to the hospital, a CAT scan revealed severe head trauma. Dr. Jim Hinsdale did not hold out much hope to Tony and Shirley that, in the event Brandon did survive, he'd lead a normal life. Brandon underwent surgery to relieve pressure on his brain, then he was moved to intensive care in a coma. His friends and family visited every day to talk to him and pray for him, but Brandon showed little improvement. He was able to open his eyes, but he remained motionless and could not speak.

After two and a half months, Brandon was transferred to The Greenery Rehabilitation Center, where an enthusiastic team of therapists, including Dr. Dennis Gallegos, worked hard to retrain him in virtually everything, from the simplest of tasks, such as swallowing, to the more difficult ones of walking and talking. After two years, Brandon was finally able to go home, still brain injured but dramatically improved.

"Brandon is courageous, truly a hero," says Gallegos. "He struggled and fought to regain control over his life. He's got a lot of guts."

Today, Brandon is attending a community college and works part-time at an amusement park food stand. He also lectures local high school students about the dangers of drinking and driving.

"I had the world at my fingertips," he tells students. "Then, in one second, it changed forever because of some stupid choice I made. You have to take responsibility for the choices you make."

Tony and Shirley are encouraged by Brandon's progress but hope that someday he will be totally self-sufficient.

"Brandon's really overcome so many things," says Shirley. "And I'm so grateful every day he's still with us."

Nail-Gun Heart

On February 22, 1992, in El Cajon, California, friends and fellow construction workers Jim Robertson and Kirk Crossman were using a nail-gun power tool to nail boards to a ceiling on a job site. The men were standing on a scaffold, and Kirk was holding his nail-gun, when Jim bent down to pick up a board and accidently bumped Kirk from behind. Kirk lost his balance, and, as he tried to regain it, the nail-gun in his hand hit Jim and discharged a nail into Jim's chest.

Kirk heard the nail-gun fire and turned around to see Jim rip open his shirt. He saw a small hole to the left of Jim's sternum where the one-and-three-quarter-inch-long nail had entered his heart.

Instead of calling 9-1-1, Kirk and Jim jumped in Kirk's truck and sped to the nearest hospital. Jim seemed to be doing okay until a half-mile into the drive, when he complained of severe neck pain.

"I'm going to die," Jim said, then he started convulsing.

Jim's head hit the gearshift, throwing the truck into park. Kirk was frightened, knowing he had to get Jim help now. As he hit the gas pedal, he saw a Sheriff's Department car in the parking lot of a mini-mall. Kirk raced across the boulevard's median and screeched to a halt in front of Deputy Lenice Lopez. Lopez took one look at Jim, who was cold and sweating, and she knew he was in critical semiconscious condition.

Lopez's partner radioed Life Flight, but the helicopter was grounded due to fog, so the Sheriff's Department helicopter, ASTREA, was dispatched to the scene. Meanwhile, Lopez tended to Jim.

"I saw how fast he was dying," recalls Deputy Lopez, "and he needed to fight."

Lopez cradled Jim's head in her arms and encouraged him to hang in. She took his hand in hers and told him to squeeze it once for yes and twice for no.

"Do you have any children?" asked Lopez. Jim squeezed her hand once.

"How many?"

He squeezed her hand three times. Lopez told Jim he had to fight to survive for his kids because they needed their father.

Kirk stood by, emotionally distressed. He'd known Jim for several years and had always looked up to him. He was praying that things would work out well for his friend.

Lopez was relieved when rescue units from the San Miguel Fire Department arrived, followed by Hartson Ambulance. Paramedic Michael McKinley recalls that Jim's chest was purple and the veins in his neck were the size of breadsticks due to a backup of blood in his system, a classic sign of a piercing injury to the heart.

ASTREA landed on the scene and the engine was kept running so Jim could be "hot-loaded" to save crucial minutes.

"They talk about the golden hour," says McKinley, "but in this case it was more like the golden minute. He needed to be in now."

Lopez said good-bye to Jim, and he whispered, "Thank you."

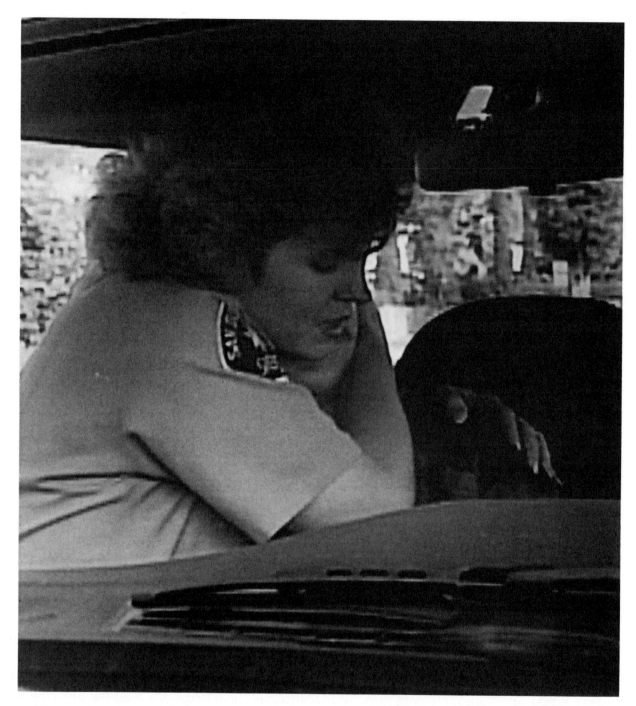

"I was so upset with him," recalls Lopez. "Thank me if I save your life, but not if I don't."

As the helicopter departed, Lopez asked McKinley if he thought Jim would live.

"It will be a miracle," he responded.

At Sharp Memorial Hospital, Jim was rushed into surgery to repair the hole in his heart. His parents arrived and were told by doctors that their son had survived, but only because the system had worked properly and swiftly.

Jim was released from the hospital just four days after surgery, but the accident has changed his life. Unable now to do strenuous construction work, he must find a new career. But he is grateful to be alive for his children's sake.

"I owe my life to everyone who was involved that day," says Jim. "The whole system worked great. These people are heroes."

Kirk says the accident has taught him a lesson.

"I'm more cautious on the job site," he says. "But if it ever does occur again, I will call 9-1-1. I won't try to be the ambulance and paramedic myself."

Jim and Lenice Lopez continue to keep in touch regularly.

"Jim and I have a bond that only people who have been in this situation will understand," says Deputy Lopez. "I think there was an angel watching over him that day."

"She gave me the mental strength to hang on," says Jim. "God sent me an angel when he sent me Lenice."

Insulin Shock Dad

On the afternoon of June 10, 1992, six-year-old Samantha Barth of Federal Way, Washington, was riding home from her day care center with her father, Nels, who was driving erratically, weaving, and making sloppy turns. Nels pulled into the driveway of the house and parked the car.

"Are you feeling low, Dad?" asked Samantha.

"I'm fine," Nels responded sluggishly.

Samantha coaxed her father out of the car and escorted him into the house. Nels said he was going to take a nap, but he paced the kitchen in a daze, then lay down on the couch. Samantha was concerned because her mother and father had taught her about her father's diabetic condition, and she knew that "feeling low" was a warning sign that he needed to eat.

Samantha sat on the couch near her father and watched television. She felt his arm to see if he was sweating, another warning sign her parents had taught her. Nels was indeed beginning to sweat, so Samantha tried to encourage him to have something to eat. Nels just groaned in response. Unwilling to take no for an answer, Samantha went into the kitchen and got an orange, a bag of potato chips, and a soda and brought them to her father's side. Nels took a sip of soda but fell back to "sleep."

When Samantha touched her father's arm again it was very sweaty. She shook him, but he wouldn't wake up. Samantha was scared and knew she had to do something.

"If he's sweaty, he's either low or high," she says. "So I just had to save his life."

When she was a toddler, Samantha's parents had taught her to call 9-1-1 in case of an emergency. Samantha was certain this was an emergency, and she dialed the number. Federal Way Fire Department Dispatcher Pat Everett answered the call.

"My daddy's low," Samantha reported to Everett. "He's a diabetic."

While Everett and another dispatcher, Vern Vandiver, took turns staying on the line with Samantha, Federal Way Fire Department rescue units and King County paramedics were dispatched to the scene.

"It's scary to be home alone and there's no one with you when your daddy's low on sugar," recalls Samantha.

Everett told Samantha that help was on the way and instructed her to shake Nels, hard. Samantha couldn't wake him, and Everett worried whether rescue units would arrive quickly enough in case Nels needed CPR.

Within three minutes, the fire department arrived with Lt. Robert Stinnett in charge. EMTs found Nels in a semiconscious state and immediately administered glucose. Moments later, the advanced life-support unit arrived with paramedic Chris Merritt and determined that Nels's blood sugar level was dangerously low. They administered an intravenous dose of a highly concentrated sugar solution, and within minutes, Nels became alert and was thanking his rescuers.

Nels, a baggage handler who had started a new shift that morning, had mistakenly attributed his grogginess to fatigue rather than low blood sugar.

The Federal Way Fire Department presented Samantha with an award for saving her father's life. All the rescuers involved offer Samantha glowing praise.

"She did a fantastic job," says paramedic Merritt. "If her parents hadn't educated her to the signs and symptoms of insulin shock, her father would have laid there and died."

"Samantha was fabulous," praises dispatcher Everett. "She saw her dad was in an obvious diabetic state. She assessed the situation and took immediate action. She's six years old! She's one in a million."

Samantha's parents are also very proud of her, although according to Nels, "Pride in my daughter? That's an understatement!"

Samantha is also very pleased she saved her father's life.

"I'm happier than anything in the whole entire world," she says, "because I saved his life and I love him 'infinity.' "

Balcony Fall

On April 10, 1992, sixteen-year-old Andrea Jones and her best friend, Rachel Koets, were enjoying their last night of spring vacation in the beach town of Gulf Shores, Alabama. After an evening of socializing, Andrea and Rachel returned to the Jones family condominium shortly before their midnight curfew. Although they said good night to Andrea's mother, Kathy, the girls secretly had other plans. They wanted to meet up with some boys they had met earlier that night.

Leaving through the front door would be too noisy, they thought, so the girls decided to climb down the balconies from the sixth-floor condo to the ground.

Recalls Rachel, "Andrea looked at me and said, 'Do you think we can scale the wall?' And instantly I thought, hey, that would be cool, because Andrea and I never did anything exciting or rebellious."

Rachel, who stands five feet ten inches, would go first in order to help Andrea, since the reach between balconies would be a more difficult stretch for her five-foot-three friend. Rachel climbed over the Jones's railing, clung to the balcony bars, and easily dropped to the fifth-floor porch below. Andrea swung herself over the railing, and Rachel successfully helped pull her down to the porch. The girls repeated their moves on the fourth-floor balcony.

Kristin Kuhn, who was watching television with a girlfriend inside her third-floor condo, was startled to see a girl land on her balcony. She watched as Rachel struggled to pull Andrea's feet down. But Andrea's legs were flailing, and she screamed that she was going to slip. Before Rachel could grab hold of her legs, Andrea lost her grip and fell twenty-five feet, from the fourth floor to the concrete below.

A call for help sent Gulf Shores fire and rescue units to the scene with Corporal Mitchell Sims and paramedic Charlie Ingram. Upon arriving, Ingram immediately radioed for a Life Flight helicopter, although he expected Andrea to be pronounced dead on arrival at the hospital. She was having a seizure and bleeding from her left ear, a sign of a closed head injury. Andrea's parents, Kathy and Jerry Jones, rushed downstairs when they heard the news.

"Words can't describe what it's like to see your child suffering and in pain," says Kathy. "All I kept thinking was, please God, take care of her and keep her alive."

There was nothing Kathy could do to help her daughter, but at least she could comfort Rachel. After searching for her, Kathy found Rachel on the beach, weeping in distress.

LifeFlight airlifted Andrea to Baptist Hospital where a trauma team, led by Dr. James Leker, met her arrival. Tests and X-rays revealed that, amazingly, Andrea had no broken bones—she had landed directly on her head. Dr. Leker felt her head injury was so severe she probably wouldn't survive. He discussed the possibility of organ donation with Kathy and Jerry, and only then did they realize how critical their daughter's condition was.

Over the next twelve hours, Andrea continued to deteriorate as her brain swelled. In a last-ditch effort to save her, doctors performed an unusual emergency procedure, removing a sizable portion of her skull to give the brain room to swell and heal.

Andrea remained in a coma for one week, then awakened on Good Friday. By Easter, she was removed from the respirator and soon she returned home to Grand Rapids, Michigan, where she underwent physical therapy and surgery to repair her skull.

Miraculously, Andrea suffered no permanent neurological damage as a result of the accident, and today is a cheerleader at school.

Andrea and Rachel are not as close friends as they were before the accident. Andrea wishes they were, but understands that Rachel stays away because she feels so guilty.

"I've learned to think before I do things," says Andrea. "By not thinking, you're hurting people that really love you and you're hurting yourself."

Andrea's parents feel very lucky to have their daughter back.

"Everyday I cherish that I can see her and talk to her," says Kathy, "and know she's still part of our family."

Two-Year-Old Pool Save

The Bradley family was excited about their new home in Englewood, Colorado. It had a pool in the backyard and the children loved to swim.

On June 30, 1992, a few days after moving in, Rosemary Aird-Bradley promised eight-year-old Kristen, five-year-old David, and two-year-old Stephen that they could go swimming after she and her in-laws, Donald and Estelle Bradley, finished unpacking boxes. So the kids put on their bathing suits, helped unpack, and explored the new house.

Stephen wandered into the kitchen. He noticed the sliding glass door leading to the backyard was open, so he toddled outside.

As Rosemary unpacked boxes in the basement, she realized she had neither seen nor heard from Stephen in over a minute. Concerned about the pool, which was not yet

fenced, she rushed outside. Rosemary couldn't see Stephen because he was sitting under a bench against the wall to her right. After scanning the pool and the yard, Rosemary went back inside.

"Okay, everybody look for Stephen," Rosemary told her family.

Rosemary, Estelle, and the kids searched upstairs, downstairs, and in the basement. Kristen recalled seeing the open sliding glass door, so she went outside. To her horror, she saw Stephen's small body floating face down in the pool.

Kristen jumped into the water and screamed to her mother as she swam to Stephen and carried his lifeless body out of the pool. Rosemary heard Kristen's scream and ran outside. She helped Kristen lay Stephen on the ground and shouted to Estelle to call 9-1-1.

Dispatcher Doug Terry, of South-Metro Fire Dispatch, answered Estelle's call.

"We have a two-year-old, and he's split his head open. He fell in the pool and we need help," Estelle reported, trying to remain calm. As she looked across the yard, Estelle was confusing Rosemary's red blouse for blood. She didn't realize that the problem was much more serious, nor did dispatcher Terry, at first.

Estelle ran outside with the phone.

"I think he's dead," Rosemary said to her in tears. "He's not breathing." Stephen didn't have a pulse, either. As Rosemary, a nurse, racked her brain to remember her CPR training, she felt devastated and overwhelmed with guilt that she could let something like this happen.

"We think he's dead," Estelle relayed to Terry, struggling not to fall apart.

"Is he breathing?" asked Terry.

"He's not breathing."

Terry flipped open his medical cards and told Estelle to relay his exact instructions to Rosemary. As Terry walked them through CPR over the phone, the Arapahoe County Sheriff's Department and Castlewood Fire Department were dispatched to the scene. Deputy Teresa Keenan was first to arrive, four minutes later.

Deputy Keenan took over administering CPR until the advanced life-support unit arrived, led by paramedic Bob Smith. Smith raced to Stephen, who was still without breath or pulse, picked him up, and ran to the ambulance. Paramedics continued to administer CPR en route to the hospital, and soon they detected a heartbeat.

Stephen was treated at Littleton Hospital by Dr. John Riccio. When he began to "posture"—or become very combative, a sign of serious brain injury—Dr. Riccio worried that if Stephen did survive, he would have severe brain damage.

Stephen was airlifted to Porter Memorial Hospital in a comatose state. Two days later, he awakened. Remarkably, the next morning, he was running down the halls. He was released from the hospital with no signs of physical or neurological damage.

Today, the Bradleys' pool is fenced and padlocked, something they should have done, says Rosemary, before they moved in. The family is overwhelmingly grateful that Stephen was given a second chance.

"You find yourself realizing how precious life is and how quickly it can be snatched away," says Stephen's father, Barrie. "So do your best to appreciate everything that life has to give you."

Maui Sand Trap

On the morning of August 18, 1992, on the Hawaiian island of Maui, Angela Campos waved good-bye to her five-year-old daughter, Sandra, and her husband, Joseph, who were leaving to go fishing while Angela got ready for work.

Joseph was on the highway driving Sandra in his sub-compact car toward their fishing spot when he was forced to stop near an intersection. The car in front of him had just stopped suddenly to avoid hitting another car. At the same time, a semi-trailer truck hauling over twenty tons of sand was approaching the intersection that was perpendicular to Joseph. The truck driver saw the cars stopped in his path in the middle of the intersection. He applied the brakes but realized he wasn't going to be able to stop in time—so he instinctively made a right turn. The trucker avoided hitting the cars, but in the process his trailer overturned.

Joseph saw the approaching truck start to flip over. He pushed Sandra down on the seat and threw himself over her. A second later, the trailer landed on top of his car, burying it under tons of sand.

A passing motorist, J. R. Hawaii, stopped and ran to the overturned trailer. He was shocked to hear Joseph's and Sandra's cries for help, because he couldn't imagine anyone would be alive inside a car crushed under twenty tons of sand. J. R. started digging with his hands and yelled to other bystanders to do the same.

Another motorist, Janet Bostick, also stopped to help.

"I asked, 'Is anybody alive?' " recalls Janet. "And they said, yeah. And I'm looking and I'm thinking, you're kidding. How can anyone be alive in this?"

Janet figured she was digging to uncover dead bodies, until she, too, heard Joseph calling out that he needed air.

The truck driver called 9-1-1 and within seven minutes, an International Life-Support Rescue Unit arrived with paramedic Mark Schafer. But there was little rescuers could do for Joseph or Sandra until the trailer was lifted off the car. Schafer thought the odds were slim that a rescue could occur in time for either one to survive.

Maui Fire Captain Conrad Ventura, who arrived moments after the fire department unit, realized that heavy equipment was needed to aid in the rescue. He radioed for a crane, which arrived ten minutes later. As the crowd watched, the trailer was attached to the crane's cable and slowly lifted off the car.

Schafer was now able to get to Sandra, who had been buried under the trailer for one hour. He quickly stabilized her and transported her to the hospital. Extricating Joseph was more difficult, but rescuers finally reached him after cutting the roof off his car. After Joseph was also pulled to safety, the crowd of bystanders burst into applause, delighted and amazed that both victims were still alive.

Miraculously, Sandra suffered only cuts and bruises and was released from the hospital the same day. Joseph sustained compressed fractures in his back and nerve damage to his right side. He had saved Sandra and himself by throwing his body on top of her.

Coincidentally, Angela Campos had driven past the accident on her way to work and had no idea that its victims were her husband and daughter. She is very thankful that they both survived.

"I am so grateful to those people who dug them out," says Angela. "If I could be face to face with every one of them, I would say thank you so much for taking care of my husband and daughter, people you didn't even know, and weren't even sure were alive."

Sandra is also very thankful, especially to her father.

"To me he's a hero."

Escalator Traps Boy

On January 20, 1993, Stephanie MacIntosh and her three children were waiting for a train in the Rundle Light Rail Transit station in Calgary, Alberta, Canada. Stephanie told her children to stay close by while she made a call at a pay phone. Five-year-old Rebecca and eighteen-month-old Taylor stood right by her side. But Michael, an active four-year-old, a real handful according to his mother, wandered away.

Michael hopped on a nearby escalator, sat on a step and rode down. When he reached the bottom, the sleeve of his jacket got caught in the floor plates and he was pulled to the ground. As the stairs continued to move, the metal plates sucked up Michael's clothes, and within seconds, his jacket and sweatshirt became tightly wrapped around his neck. Michael screamed in pain.

A commuter, eighteen-year-old Nathan Gray, stepped on the escalator and heard a wail like nothing he'd ever heard before. He looked down and saw a child lying at the bottom of the stairs. Nathan ran down and tried to yank Michael free, but he couldn't move him an inch. The escalator was cinching his clothes tighter and tighter, and now his arm was caught in the plates—and losing circulation fast. Still worse, the clothes wrapped around his neck were beginning to strangle him.

"Shut it off!" shouted Nathan as he scrambled to find the escalator's emergency stop button.

Stephanie heard Michael's screams and ran down the escalator. Michael's breathing had become shallower. His screams were growing fainter. Then he became quiet.

"My heart was being ripped apart," remembered Stephanie as she watched his face grow deep purple. "He just went into a cold stare, and I just kept telling him, 'Michael, Mommy's here.' "

A passerby finally pressed the emergency stop button, but Michael's clothes continued to strangle him. Nathan whipped out his pocket knife to cut off the clothes, but they were wrapped so tightly around Michael's body that Nathan could barely slip his knife under them.

Meanwhile, a concession stand employee had called 9-1-1 and was speaking with dispatcher Don Papineau, but the caller's heavy accent prevented Papineau from understanding what the emergency was. When another commuter, James Hilson, stopped to help Michael, Nathan handed him the knife and told him to keep cutting. Then Nathan ran to talk to the dispatcher.

"Is he breathing?" asked Papineau.

"Not from what we can tell."

Papineau asked Nathan to check on Michael's breathing and report back to him. By the time Nathan arrived at the escalator, Michael had begun screaming again.

"It was the most beautiful sound I'd ever heard," recalls Nathan.

Rescuers from the Calgary Fire Department arrived on the scene, including paramedic Doug Lewis, who couldn't believe his eyes. How could Michael possibly have become trapped like that, Lewis wondered. It looked as if part of his body had been pulled into

the escalator plates. His right arm had become severely twisted and was turning blue from loss of blood circulation. His face was covered with small blood spots, indicating intense suffocation.

In order to free Michael, rescuers needed to lift off the metal plates. There was so much clothing and skin trapped under them, it was difficult to loosen the screws. When the plates had been removed, rescuers got a good look at Michael. His right arm was broken and he had pinch marks from his shoulder down to his hips. Lewis felt that the boy was lucky to have escaped with such relatively minor injuries considering what he'd been through.

"Nathan is eighteen years old," says Stephanie, "and he's got a good head on his shoulders for doing what he did. I couldn't express enough thanks to him. If he hadn't been there, Michael probably would have died."

Paramedic Lewis says parents need to realize that escalators pose a danger to children. He advises that when an escalator is nearby, parents take their children by the hand and make sure their clothing isn't dangling.

Michael also has some good advice for children.

"You should stay where your mama is. And daddy."

"Michael's a junior Evel Knievel," says Stephanie. "He's not afraid to do anything. Except he doesn't like to go near an escalator now unless he's right beside me holding my hand."

Ex-RN Son Save

On the night of June 24, 1987, in Knoxville, Tennessee, nineteen-year-old Trent Winston and his two buddies were intoxicated as they drove home from an evening on the town. The driver, Trent's friend, lost control of his car and skidded off the road. The car hit a tree, flipped over, and landed on its roof. The boys were trapped inside. In the front passenger seat, Trent was suspended by his seat belt upside down, his face crushed against the roof of the car.

A neighbor heard the crash and called 9-1-1. A few blocks away, Donna Davis had just come home from work when she also heard the crash. Donna, a former emergency room and intensive care nurse, hopped back in her car and drove to the scene.

Donna approached the wreckage and saw that the driver was unconscious. Another boy was screaming in the backseat. Donna took the driver's hand in hers and calmly spoke to him until he regained consciousness a few minutes later.

"Oh, my buddy," the driver moaned, "my buddy next to me's hurt."

"Are you saying there's another boy in this car?" Donna asked, surprised. She had seen only two kids in the car.

Donna ran to the passenger side. It was too dark to see clearly inside, but she heard Trent gurgling as he gasped for air. Donna knew the boy was in serious trouble because the sound meant he didn't have a clear airway.

God, I feel sorry for this kid's parents, she thought to herself, because it'll be a miracle if he lives.

Donna reached inside, found the boy's mouth, and did her best to open the airway that was filled with blood.

"I was so scared he was going to die," recalls Donna. "My heart was pounding so hard because there were so many times when he quit breathing, and I kept talking to him and I'd say, look, you are too young to die. Just hang in there and we'll get you out of here."

Within minutes, the Knoxville Volunteer Rescue Squad, led by Captain John Yu, arrived on the scene. Using the "jaws of life," Yu and his men pried open the car and extricated Trent's friends. But getting Trent out safely took longer because he was precariously situated.

Meanwhile, Donna's eighteen-year-old son, Brett, had been returning home for the evening when he heard the sirens. Concerned because his mother was not home, Brett drove to the scene. He arrived to find Donna holding a crash victim's hand in the demolished car while rescuers worked to free him.

When Trent was finally extricated from the wreckage, Donna stood back and watched rescuers load him onto a backboard and carry him to the awaiting ambulance. Suddenly, she froze. She had recognized the clothing on the boy she had helped save—he was her own son!

"I thought, this is a bad dream," Donna recalls. "This cannot be happening. Then the thought was, can we get him to the hospital alive? I told Brett, it's very bad, your brother may not live."

Trent's friends were not seriously injured, but Trent was admitted to the hospital in a coma, with wrist, leg, and jaw fractures. Doctors were concerned whether he'd recover from the head injury he had sustained. Over the next twelve days, Donna, Brett, and Trent's father—with whom he lived—stayed by Trent's side, talking to him and praying. On July 4th, Trent opened his eyes.

Five and a half weeks after awakening from his coma, Trent was transferred to the Patricia Neal Rehabilitation Center, where he spent four months relearning how to walk and talk. Today, although Trent still struggles with his speech and movements, he is able to run, water-ski, and attend college.

Trent credits his mother for motivating him to push hard in his recovery. Donna says that in her experience, patients who receive a lot of love and support from their loved ones always seem to fare better.

"I'm proud of the job I'm doing," says Trent. "The way I think is there's nothing I can't do until I've failed at it."

Trent has learned an important lesson the hard way.

"Drinking and driving don't mix," he says. "Drinking and driving kills. A lot of people have to learn that the hard way. A lot of times, the hard way makes you dead."

Aussie Snakebite Save

"We've got the top eleven deadly snakes in the world here in Australia," says snake expert Ted Mertens. "And the most dangerous snake in Australia is definitely the taipan."

On August 31, 1991, sixty-three-year-old Clive Brady enjoyed an afternoon swim in the Barron River, near his home in Mareeba, Queensland, in northern Australia. After his swim, as Clive walked along the bank of the river toward home, a snake leaped from the ground and bit him several times in the leg. Clive fell to the ground, immediately overcome by dizziness, blurred vision, and nausea.

Clive thought he had been bitten by a taipan, whose unique venom contains a detergent factor designed to move its poison through the blood system in a matter of thirty to sixty seconds.

"For average Joe," says Mertens, "ten minutes to a half hour without first aid and they're cactus."

Brian Eakin and his family were spending the day at the Barron and happened to be across the river when they heard someone call out for help. Brian and his teenage son ran across the water to Clive. Brian realized Clive needed immediate medical attention, so he left his son with him and ran to find help. By a stroke of luck, the first building Brian encountered was a local hospital.

While the head nurse called for an ambulance, nurse trainee Bernie Jo Tonon, figuring the patient wouldn't be suffering from anything all that serious, grabbed some bandages and ran with Brian to Clive.

"I was glad she came with me," recalls Brian. "It was a great relief to know there was professional help."

"I didn't have any experience as a nurse," says Bernie Jo. "As a training nurse, it's just making beds."

Bernie Jo was shocked to discover that Clive was indeed in dire straits, and that his condition was deteriorating by the second. He was sweating profusely and his breathing was labored. Bernie Jo tried to keep Clive conscious by talking to him, but before the ambulance arrived, he lost consciousness.

Within moments, paramedic Steve Qazim was on the scene, and with Brian's help, carried a stretcher and medical equipment across the river. Clive was loaded on the stretcher and the group carried him to the opposite bank, where he took another turn for the worse. Bernie Joe noticed that Clive had turned blue and had stopped breathing. While Qazim retrieved oxygen from the ambulance, Bernie Jo administered rescue breathing. A moment later, Clive's heart stopped beating. Bernie Jo began CPR, which quickly got his heart started again.

"I was racing to beat the clock." recalls paramedic Qazim. "Clive was running out of time. This was very obvious to me."

As the ambulance departed with Clive, Brian's son mentioned that Mr. Brady had left his hat on the other side of the river.

Replied Brian, "Where Mr. Brady's going, I don't think he'll need a hat anymore."

At the hospital, Clive was rushed into the emergency room, and his wife and family arrived to await news of his condition. A blood sample was taken from Clive, but doctors could not afford to wait for the results. He was administered an anti-venin that was used as an antidote for all Australian poisonous snakebites.

"I knew by then that Clive wouldn't have a good chance of making it," recalls Bernie Jo. "It felt like a leaden heart to go in and see his wife sitting there. I didn't know what to say to the lady."

Results of Clive's blood sample revealed that he was in fact bitten by a taipan snake. He now was administered a specific anti-venin to counteract taipan venom and was hospitalized for six days. Remarkably, Clive did survive the snakebites and recovered without any long-term side effects. However, Clive did suffer one temporary side effect. The venom caused the hair on his leg to fall out. Six months later, it grew back thicker than ever.

Says Clive, who's bald, "Some of me mates said, 'Well, next time get him to bite you on the head. It'll put a bit there, too.' "

Friends tell Clive he survived because God didn't want him yet. But Clive, who says he's not a "religious bloke," doesn't buy that argument.

"It was the likes of Bernie Jo and the nurses and the doctors. They're the ones that helped me. If God's such a great fellow, why'd he let a bloody snake like that bite me?"

Bernie Jo doesn't think she did anything special, just what she was trained to do.

Says Bernie Jo, "I think Clive was born under a lucky star."

Fire Ant Trauma

On April 7, 1991, the Mosely family was enjoying a quiet, relaxing Sunday at their home outside Macon, Georgia. Ken watched television with a friend while Leslie washed the car outside and kept an eye on three-year-old Daniel and sixteen-month-old Toni. Daniel sat in a swing while Toni wandered around the yard, sipping juice. Then, Toni tripped.

Leslie heard Toni let out a whimper. She looked up from washing the car and saw her daughter facedown on a large hill of fire ants. Leslie ran to Toni, swept her up, and brushed the ants off her arms and legs. She noticed over a dozen ant bites on Toni, but she wasn't too concerned because Toni had been bitten over fifty times before.

Leslie set Toni in a swing and asked Ken to come outside and watch her while Leslie finished washing the car. Ken came right out and noticed Toni's eyes and lips were swollen, and her face was covered with white blotches. Figuring it must be a reaction to the ant bites, Leslie went inside to get Toni an antihistamine. Within another minute, Toni started gasping for air. Ken grabbed her from the swing and ran to the house. By the time he got inside, Toni began having seizures.

Leslie told Ken to jump in the car with Toni and drive toward the hospital. Then she called 9-1-1 and told the dispatcher Ken's route and asked him to have an ambulance intercept the car. Leslie knew that to wait for an ambulance to arrive at the house, which was located in such a remote area, might prove fatal.

"The toughest part," she recalls, "was letting Ken go with her and not knowing if she'd be alive when I got to her."

Paramedics Keith Soles and his partner, Jim Walsh, were dispatched to intercept Ken's car. They had no idea where they might encounter it, but the men were aware that people frequently die from severe allergic reactions. They hoped they could intercept Toni before her airway swelled to the point where it closed and prevented her from breathing altogether.

Ken tried to maintain his calm as he sped to the hospital with one hand on the steering wheel and the other on Toni, whom he hoped would stay conscious. As he turned off the two-lane road onto the divided interstate, Toni's condition continued to deteriorate. Meanwhile, Leslie headed to the hospital with Ken's friend.

"All I could think of on the way to the hospital," recalls Leslie, "was, God, please don't take my baby. You can never replace a child."

Ken heard a siren and pulled onto the shoulder. He jumped out of the car and hoped the paramedics, who were approaching from the opposite direction, would see him waving. Soles and Walsh did see Ken on the other side of the highway holding a baby, but rather than drive across the median, they exited the next off-ramp and reentered on Ken's side. The paramedics whisked Toni from Ken's arms, loaded her into the ambulance, and rushed to the hospital.

"The child was trying to cry," recalls Soles, "but it was in a respiratory state where it couldn't. It had to concentrate its total effort on breathing."

At Coliseum Hospital, Toni was examined and treated by Dr. Tim Graves.

"Her condition was at the breakpoint," recalls Dr. Graves. "She was either going to get better or get worse fairly quickly."

Toni was given oxygen to aid her breathing and an injection of epinephrine to reverse the effects of allergic reaction. Her condition immediately improved, and she began screaming and crying, a positive sign that she was responding to the injection. Toni was released from the hospital, a happy, healthy baby.

"Toni is back to normal," says Ken, "and she can do everything normal kids do, as well as wrap Daddy around her little finger. I'm very grateful to the emergency room people and the paramedics."

Dr. Graves feels that the Moseleys' decision to meet the ambulance on the road—unwise and not advisable in general—proved to be a smart move in Toni's case because allergic reactions can cause devastating results in minutes. He says that if Toni is stung by a bee or wasp in the future, or bitten by ants again, she could suffer an even more deadly allergic reaction the next time. For this reason, the Moseleys now carry a medical kit with epinephrine—the same drug used at the hospital—at all times.

Says Leslie, "Toni should grow up okay. She's playful and spunky, and smart and wild. And she gives her brother a hard time."

Box Canyon Rescue

On February 14, 1991, seventeen-year-old Christina Sousa and her girlfriend decided to cut school and go hiking in Box Canyon, a rugged, scenic area in Mt. Shasta, California. The girls found a trail and started down the treacherous canyon, where one-hundred-foot vertical walls of rock rise from the riverbed below. Chrissy and her friend reached some large boulders that interrupted the trail. Chrissy decided to climb across the boulders. She hugged a rock and stepped over to the next one, but lost her footing and slipped off. Chrissy fell forty feet down the shear rock wall.

Chrissy's friend couldn't see where she had landed. She yelled to Chrissy and finally heard her moaning. Terrified, the friend scrambled down the canyon and found Chrissy lying on the ground, bleeding at the back of her head. The friend said she was going to get help, but Chrissy was scared and didn't want her to leave. The friend gave Chrissy her heart-shaped necklace and promised she would return.

"Keep looking at the heart," Chrissy's friend told her. "You'll be okay. I'll be right back."

The friend climbed out of the canyon and was running down the road looking for help when she encountered classmate Gus Orosz and his friend. They all ran back to the canyon, and Gus hiked down to Chrissy. He realized he couldn't move her, so he yelled to his friend to call 9-1-1. Chrissy's friend stood at the top of the canyon to show rescuers where to go down while Gus waited with Chrissy.

Within five minutes, an ambulance and a unit from the Mt. Shasta Fire Department, trained in mountain rescue, arrived at the rim of the canyon. Paramedic Angelo Banos climbed down to Chrissy. Concerned about a possible spinal injury, and not wanting to risk moving her, he requested a helicopter.

Volunteer firefighter Phil Sousa was at work when he received word of the accident over his pager. He rushed to the scene, but was stopped by the fire chief, who didn't want Phil down in the canyon. He told Phil he thought the girl who'd fallen was his daughter. The news hit Phil hard.

"Suddenly, you have them tell you they think it's your kid," recalls Phil, "in one of the worst situations you've witnessed in eighteen years in the department. It's a real scary situation."

But Phil was determined to climb down and help with the rescue. He sent a friend to tell his wife, Jan, what had happened. Coincidentally, Jan was listening to news of the accident over the radio when the friend arrived at the house.

" 'This call you're listening to, we think it's Chrissy in the canyon,' he told me," recalls Jan. "It was almost unbelievable. I couldn't picture it. It's like I denied it really was her."

When Phil reached Chrissy, he was greatly relieved to see she was still in one piece. It was decided that a vertical rescue using ropes to bring Chrissy up the rock face in a litter was too risky for both firefighters and Chrissy. She would have to be evacuated by helicopter.

Pilot Bruce Riecke surveyed the scene from the air. The accident had occurred in the worst possible place in the canyon and Riecke could not see a landing site.

"When you first look down there," he recalls, "you can't believe someone could fall down there and survive."

Finally, Riecke thought he saw a landing site and lowered the helicopter down to a shallow part of the river at the bottom of the narrow canyon. Jan had arrived and watched from above as the helicopter departed with her daughter.

"I felt this giant weight of relief," recalls Jan. "It was a very stirring thing to see them do this thing for my daughter. I'm not religious, but I felt as if angels had come and carried her off."

Chrissy was treated for a fractured skull, broken leg, and numerous cuts and bruises. Miraculously, she did not suffer a spinal injury.

"I think it's a miracle that she ever came out of the canyon," says Phil. "It makes you feel how precious life is. It could have been a real disaster."

"I'm real glad my dad went down there," says Chrissy. "It's real difficult for him and I'm sorry he had to go through that, but I'm glad he helped me out."

Good Samaritan Save

On October 17, 1991, seventeen-year-old Jamie Scavona and her best friend, Heather Walter, were returning from a local mall to their home in Stillwater, Oklahoma. Jamie and Heather were driving down the four-lane interstate in Jamie's pickup truck, and neither girl was wearing a seat belt. Jamie was driving fast and not paying attention to the road. The girls were chatting and listening to the radio when Jamie's truck veered onto the grassy median. Jamie steered to get back on the road, but she overcompensated and lost control. Her truck swerved off the road and flipped several times before landing on the driver's side.

Heather lay in the passenger seat, bruised and badly shaken. As she came to her senses, she saw that the driver's seat was empty and that a bloodied arm was waving to her outside the window. Apparently, during the crash, Jamie had been thrown from the truck. After flipping, the vehicle had landed on her. Heather climbed out the passenger-seat window and screamed for help.

Mark Gardner happened to be driving down the interstate on his way to catch a plane when he saw Jamie's truck and heard Heather's cry for help. He pulled his car onto the shoulder of the road and ran to the scene. The sight of Jamie pinned under the truck scared him because he could picture his daughter in her place. Jamie looked pale and was barely breathing. Mark knew he had to get the truck off her. He squatted, and, through sheer adrenaline, lifted the pickup and rolled it back on its wheels.

One of the many passing motorists who stopped to look at the accident called the Oklahoma Highway Patrol from a car phone. Mark knelt by Jamie and told her to hold on and relax. Help was on the way. But Jamie, whose respiration problems were complicated by asthma, stopped breathing. Mark didn't know what to do. He asked bystanders whether anyone knew first aid. No one came forward.

"Looks like she's gone," someone said.

"No she's not, because, by God, I'm not going to let her go," responded Mark.

Mark did the best thing he could. He'd seen someone perform rescue breathing on the television program *Rescue 911*, and he decided it was better to try the technique than do nothing. He pinched Jamie's nose and breathed into her mouth several times. Jamie gasped for air and started breathing weakly on her own. Someone told Mark to give Jamie her inhaler, which helped her breathe during an asthma attack, but Mark was afraid to use it.

When State Trooper Robert Park responded to the scene, Mark asked if he would radio rescue units en route from the Guthrie Fire Department and find out if it would be safe for Jamie to use her inhaler. EMTs Mike O'Connor and Chuck Burtcher radioed back that it would be okay.

Mark finally remembered he had a plane to catch. He gave the officer his name, address, and phone number, then drove away.

Medics rushed Jamie to Stillwater Medical Center where she was treated by Dr. Doug Wilsey. She was in severe respiratory distress and suffering from acute asthma. Jamie

underwent surgery to repair a partial collapse of one lung, several fractures, including her pelvis and collarbone, and a ruptured bladder. Two and a half weeks later, she was released from the hospital.

Today, despite her asthma, Jamie is in good health. She feels she's been given a second chance at life as well as some important lessons. She doesn't speed anymore, and she wears her seat belt.

Mark and Jamie have become buddies as a result of the accident.

"I owe everything to Mark," says Jamie. "If he hadn't shown up, I wouldn't be here today. To me he's a hero."

Mark says he also learned some lessons. He enrolled in a first-aid and CPR class. And he learned it pays to get involved.

"Don't be afraid to stop and help," he says. "You *can* make a difference."

**If you enjoyed *Rescue 911: Extraordinary Stories*
you may want to order these other exciting
Rescue 911™ Products:**

ITEM No.	TITLE	PRICE
0681452587	Rescue 911 1994 Safety Tip-A-Day Calendar	7.95
0681452579	Rescue 911 1995 Wall Calendar	8.99

Ordering is easy and convenient.
Order by phone with Visa, MasterCard, American Express or Discover:
☎ 1-800-322-2000, Dept. 706
or send your order to:
Longmeadow Press, Order/Dept. 706,
P.O. Box 305188, Nashville, TN 37230-5188

Name _____

Address _____

City _____ State _____ Zip _____

Item No.	Title	Qty	Total

Check or Money Order enclosed Payable to Longmeadow Press

Charge: ☐ MasterCard ☐ VISA ☐ American Express ☐ Discover

Account Number

☐☐☐☐ ☐☐☐☐ ☐☐☐☐ ☐☐☐☐

Subtotal	
Tax	
Shipping	2.95
Total	

Card Expires

☐☐☐☐ Signature _____ Date _____

Please add your applicable sales tax: AK, DE, MT, NH, OR, 0.0% — CO, 3.8% — AL, HI, LA, MI, WY, 4.0% — VA, 4.5% — GA, IA, ID, IN, MA, MD, ME, OH, SC, SD, VT, WI, 5.0% — AR, AZ, 5.5% — MO, 5.725% — KS, 5.9% — CT, DC, FL, KY, NC, ND, NE, NJ, PA, WY, 6.0% — IL, NM, UT, 6.25% — MN, 6.5% — MS, NV, NY, RI, 7.0% — CA, TX, 7.25% — OK, 7.5% — WA, 7.8% — TN, 8.25%